AF781436

Mobility and Technology in the Workplace

The contemporary period has witnessed a rapid evolution in a wide range of mobile technology. This book charts the profound implications these technological changes have for workers and business organizations. From an organizational point of view they have the potential to transform the nature of organizations, by enabling workers to be increasingly mobile. From the perspective of workers these changes have the potential to impact on their work-related communications, how they manage the increasingly blurred public–private divide, and the nature of the home–work boundary.

These chapters afford a detailed insight into these issues by bringing together an international collection of contemporary studies and analysis and taking a critical perspective towards some of the advertised myths regarding mobile technology usage. Issues covered include:

- travel and the changing nature of spatial mobility patterns;
- work–space and place and the 'leaking' out of organizations into more public domains;
- mobile work practices, including detailed and heterogeneous case studies;
- home–work dynamics and the changing nature of the home–work boundary;
- implications for Public Policy.

This book will be of great interest to researchers engaged with business, geography and information systems, as well as students on Management Studies MBAs and MScs.

Donald Hislop is a Senior Lecturer at Loughborough University Business School.

RIOT!

Routledge studies in innovation, organization and technology

1 **Innovation in the U.S. Service Sector**
Michael P. Gallaher, Albert N. Link and Jeffrey E. Petrusa

2 **Information and Communications Technologies in Society**
E-living in a digital Europe
Edited by Ben Anderson, Malcolm Brynin and Yoel Raban

3 **The Innovative Bureaucracy**
Bureaucracy in an age of fluidity
Alexander Styhre

4 **Innovations and Institutions**
An institutional perspective on the innovative efforts of banks and insurance companies
Patrick Vermeulen and Jorg Raab

5 **Knowledge and Innovation in Business and Industry**
The importance of using others
Edited by Håkan Håkansson and Alexandra Waluszewski

6 **Knowledge and Innovation**
A comparative study of the USA, the UK and Japan
Helen Brown

7 **Industrial Innovation in Japan**
Edited by Takuji Hara, Norio Kambayashi and Noboru Matsushima

8 **Managing and Marketing Radical Innovations**
Marketing new technology
Birgitta Sandberg

9 **Mobility and Technology in the Workplace**
Edited by Donald Hislop

166383842
12/08
CAMERON UNIVERSITY LIBRARY
2800 W. GORE BLVD.
LAWTON, OK 73505-6377
HD
58.8
.M574
2008

Mobility and Technology in the Workplace

Edited by Donald Hislop

Routledge
Taylor & Francis Group
LONDON AND NEW YORK

First published 2008
by Routledge
2 Park Square, Milton Park, Abingdon, Oxon OX14 4RN

Simultaneously published in the USA and Canada
by Routledge
270 Madison Ave, New York, NY 10016

Routledge is an imprint of the Taylor & Francis Group, an informa business

© 2008 Selection and editorial matter, Donald Hislop; individual chapters, the contributors

Typeset in Times by Wearset Ltd, Boldon, Tyne and Wear
Printed and bound in Great Britain by MPG Books Ltd, Bodmin

All rights reserved. No part of this book may be reprinted or reproduced or utilized in any form or by any electronic, mechanical, or other means, now known or hereafter invented, including photocopying and recording, or in any information storage or retrieval system, without permission in writing from the publishers.

British Library Cataloguing in Publication Data
A catalogue record for this book is available from the British Library

Library of Congress Cataloging in Publication Data
Mobility and technology in the workplace/edited by Donald Hislop.
p.cm. – (Routledge studies in innovation, organizations and technology; 9) Includes bibliographical references and index.

1. Organizational change. 2. Mobile computing. 3. Employees – Effect of technical innovations on. I. Hislop, Donald.
HD58.8.M574 2008
331.25′6–dc22 2008001050

ISBN13: 978-0-415-44346-3 (hbk)
ISBN13: 978-0-203-89435-4 (ebk)

Contents

Figures

Tables

Contributors

Carolyn Axtell, Senior Researcher, Institute of Work Psychology, University of Sheffield, UK.

Lyn Batchelor, Lecturer, University of Aberdeen Business School, Aberdeen, UK.

Professor Jonathan V. Beaverstock, School of Geography, University of Nottingham, UK.

Ann Bergman, Department of Work Life Science, Faculty of Economic Sciences, Communication and IT, Karlstad University, Sweden.

Silvia Elaluf-Calderwood, Doctoral Student, Department of Management, Information Systems and Innovation Group, The London School of Economics and Political Science, UK.

James R. Faulconbridge, Lecturer, Department of Geography, Lancaster University, UK.

Laura Forlano, Doctoral Student, Department of Communications, Columbia University, New York, USA.

Anne E. Green, Principal Research Fellow, Institute for Employment Research, University of Warwick, UK.

Per Gustafson, Post Doctoral Fellow, Institute for Housing and Urban Research, Uppsala University, Sweden.

Professor Irene Hardill, Business, Law and Social Sciences Graduate School, Nottingham Trent University, UK.

Donald Hislop, Senior Lecturer in Organizational Behaviour and HRM, Loughborough University Business School, UK.

David Holley, Research Student, Centre for Transport and Society, University of the West of England, UK.

John Holm, Researcher/Consultant, Woods Bagot.

Juliet Jain, Research Associate, Centre for Transport and Society, University of the West of England, UK.

Tommy Jensen, Assistant Professor, Department of Business Administration, Umeå School of Business, Sweden.

Professor Gavin Kendall, Humanities Research Program, Queensland University of Technology, Australia.

Diannah Lowry, Principal Lecturer, Bristol Business School, University of the West of England, UK.

Professor Glenn Lyons, Director, Centre for Transport and Society, University of the West of England, UK.

Catherine A. Middleton, Canada Research Chair, Associate Professor, School of Information Technology Management, Ted Rogers School of Management, Ryerson University, Toronto, Canada.

Megan Moskos, National Institute of Labour Studies, Flinders University, Australia.

Daniel Pica, Post Doctoral Research Fellow, Department of Management, Information Systems and Innovation Group, The London School of Economics and Political Science, UK.

Professor Colin G. Pooley, Department of Geography, Lancaster University, UK.

Carsten Sørensen, Senior Lecturer, Department of Management, Information Systems and Innovation Group, The London School of Economics and Political Science, UK.

Dr Keith Townsend, Centre for Research on Work, Organization and Wellbeing, and Department of Employment Relations, Griffith University, Australia.

Dan Wheatley, Doctoral Student, Business, Law and Social Sciences Graduate School, Nottingham Trent, UK.

Mikael Wiberg, Associate Professor, Department of Informatics, Umeå University, Sweden.

Acknowledgements

Thanks to Steve Fleetwood for providing me with the opportunity to participate in the Knowledge Economy programme at Lancaster University's Institute for Advanced Studies in July 2006, as it was while there that the original idea for the book was born.

Thanks to Catriona, Lorna and Kara for ongoing love and support.

Thanks to Julie Collett at Loughborough University Business School for administrative support in putting the final draft of the manuscript together.

1 Introduction

Donald Hislop

Context

Two relatively recent ongoing, and arguably significant ways in which work has been changing, both of which began to become visible and significant trends from the mid-1990s onwards, are first, the extent to which work is carried out in locations beyond the home and the workplace, and the increasing levels of spatial mobility required of workers, and second, the rapid evolution in a wide range of mobile information and communication technologies and their increasingly widespread use by workers. For example, mobile communication technologies such as mobile phones have developed enormously, and the start of the twenty-first century witnessed the development of mobile email devices such as BlackBerries. Personal and mobile computer technologies have also developed enormously, with laptops becoming increasingly more powerful, more portable, and more able to connect to the internet from diverse locations (through wifi). The central focus of the chapters in this edited collection is on the many and diverse ways in which these two trends have combined to change work practices and the character of the places in which work is carried out.

Such a focus connects with the burgeoning sociological literature on and interest in mobility, or mobilities, which builds from the assumption that the growing mobility of people, goods, money, ideas, etc. represents one of the defining characteristics of capitalism in the early twenty-first century (Hardill and Green 2003; Kaufmann 2002; Kellerman 2006; Larsen *et al.* 2006; Sheller and Urry 2006; Urry 2000). The contributions here add to this body of work through providing insights into mobility and mobile technology use in the domain of work, a relatively neglected focus of interest in the mobilities literature.

Often, with a hint of technological determinism, developments in mobile information and communication technologies (hereafter m-ICTs) have been argued to be instrumental in increasing the spatial mobility of work (Cooper *et al.* 2002). However, the functionality of m-ICTs is such that they simultaneously have potentially contradictory consequences for where, when and how (collaborative) work is organized. Mobile technologies such as mobile phones, BlackBerries and wifi-enabled laptops do have the potential to 'free' work/ers from the confines of fixed locations, enabling workers to be more mobile than previously, allowing work to be conducted from a greater range of locations than previously possible. Simultaneously however, m-ICTs also have the potential to reduce the need for workers to be spatially mobile, as the increasingly powerful communication and information-sharing potential of these technologies facilitates remote collaboration and arguably thus could actually reduce the need for business people to travel for face-to-face meetings (Virilio 1999). Such paradoxical, simultaneously contradictory potential implications of mobile technology use reflects what Arnold (2003) suggests is the inherently Janus-faced nature of mobile technologies/phones.[1]

Broadly, the book takes a social shaping/constructivist perspective on technology, highlighting the two-way relationship between the technical and the social through being sensitive to both the ways in which human choices in technology use shape the characteristics of technologies which facilitate mobile working and also how the use of such technologies impacts on the character of work locations, activities and experiences (Wajcman 2006). Thus the way in which m-ICTs impact on work are regarded as being as much due to negotiations between (and the choices of) social actors as being the result of the determining effect of the technologies' inherent characteristics. Thus, while the contributions in this chapter look at how both trends have impacted on work, they don't assume that it is the use of m-ICTs which is the prime determinant driving the spatial (re)organization of work.

Accompanying the dramatic and rapid developments in m-ICTs, as occurs with much technological change, has been a positive and optimistic rhetoric, or 'cyberbole' (Woolgar 2002), which suggests that these changes are both significant, and largely positive. A key component of the rhetoric surrounding the development and use of mobile communication technologies is the 'anytime, anyplace' idea (Cooper *et al.* 2002; Perry *et al.* 2001). This is articulated clearly and succinctly by Katz and Aakhus (2002, p. 303),

> The ambitious dream of a generation of telecommunications engineers is becoming realized: we will be able to have at least potential contact with most anyone at any time or place.... We can have perpetual contact, enjoying instant or asynchronous communications from around the world or the heavens above.... These are not dreams but achievable today.

The contributions in this book take a deliberately more critical, sceptical and questioning stance to such claims, examining the positive and negative consequences these changes have for workers.

Statistics

Before proceeding any further, it is helpful to empirically flesh out the claims regarding changing spatial mobility patterns and m-ICT use made above. Statistics from a range of countries illustrate the extent to which they are changing, and how an increasingly significant proportion of workers regularly need to be spatially mobile. First, Felstead *et al.* (2005), using longitudinal data taken from the UK government's Labour Force Survey showed how, between 1981 and 2002, the percentage of UK workers working 'mainly in different places using their home as a base', increased from 3.8 to 7.1 per cent of the UK's workforce, an increase of over 231 per cent in two decades. Second, Gareis *et al.* (2006, Table 3.3) found that in the 15 EU countries surveyed,[2] on average 15 per cent of these country's workforces spent at least ten hours per week working away from their home or office premises (a category of work they define as highly intensive mobile work).[3] Finally, Gustafson's (2006) study of Swedish workers, which is based on data collected between 1995 and 2001, found that 11 per cent of the workers studied had spent at least two nights out of the previous two months away from home on work-related business.

A number of studies also substantiate the claim that recent times have witnessed a significant increase in the work-related use of m-ICTs. In terms of mobile phone use, a study in the UK conducted by the London School of Economics and the Carphone Warehouse found that 46 per cent of those surveyed made use of mobile phones in their work with 15 per cent reporting that their work, would be virtually impossible without a mobile phone (LSE/Carphone Warehouse 2006). Haddon and Brynin's (2005) study of telework across a range of European countries including the UK, Italy, Germany, Norway, Israel and Bulgaria, found that the category of workers they defined as 'mobile users' ('people who say their mobile phone is important for their work but are not internet or PC home users'), accounted for on average 20 per cent of respondents. Without providing much specific empirical detail, Middleton and Cukier (2006) suggest that the work-related use of BlackBerry mobile email devices in Canada and the USA grew rapidly during the early years of the twenty-first century with RIM, the manufacturer of BlackBerries boasting over four million subscribers in 2005.

Finally, Gareis *et al.*'s (2006) research shows that, in only three years, between 1999 and 2002, mobile eWork (work which involves both spatial mobility and the use of an online computer connection when mobile) across the 15 EU countries that they studied increased from accounting for 1.5 to 4 per cent of employment. The relatively limited numbers reported in this category of work relates to the rather narrow and specific way this category of work is defined, which excludes workers who either use mobile phones, or laptops without online connections. Arguably, if eWork was defined in broader terms, the number of workers so categorized may be significantly higher.

Overall, therefore, available evidence indicates that, not only do an increasingly significant proportion of workers need to be spatially mobile to carry out their work, but that the extent to which m-ICTs are used by workers is also increasing significantly.

While there is a significant overlap and interrelationship between the need for workers to be spatially mobile, and their use of m-ICTs, it is important to acknowledge that the two are not inseparable. Not only is it possible for people to use m-ICTs without their work demanding great spatial mobility (as was the case with the office and home based workers studied by Towers *et al.* 2006), work can also involve significant levels of spatial mobility without inevitably requiring the use of m-ICTs (as is the case with postal delivery staff). However, the contributions in this book typically focus on workers who need both to be spatially mobile, and to make regular use of mobile computer and communication technologies.[4]

Definitions and terminology

A plethora of labels exists to categorize the type of work examined here. Some of the most common labels for such forms of work are nomadic or multi-location working, mobile eWork (Gareis *et al.* 2006), mobile virtual work (Andriessen and Vartianinen 2006), and mobile telework (Brodt and Verburg 2007; Daniels *et al.* 2001; Hislop and Axtell 2007). As the contributors to this book come from a wide range of countries and academic disciplines, no attempt is made to develop, or impose a unitary definition, with a catholic approach to terminology being preferred. If the academic telework literature, which is significantly more mature than the literature on mobile working, is taken as a guide, such definitional pluralism may continue and attempts to develop a wide consensus on an agreed lexicon of key terms may prove futile (Sullivan 2003; Haddon and Brynin 2005). What is arguably more important than the particular labels people employ is that everyone clearly articulates the definitions of the terms they use, which allows effective comparisons to be made. The preference in this introduction is for the term mobile telework. However, this is not intended to privilege this term over any of the others that can be used to categorize this form of work.

While this introduction has made clear that the focus in the book is on workers who are both spatially mobile, and employ m-ICTs in their work, both the amount and type of spatial mobility, as well as the types of technology used and the extent to which they are used has been deliberately left vague and ambiguous, through use of terms such as ‘regular’ and ‘significant’. While the use of such terms can be criticized for a lack of precision, the advantage of adopting such deliberately open terminology is that it avoids the risk of imposing a somewhat arbitrary boundary on terms (such as: to be classified as a mobile worker, someone must spend at least two days per week working away from any home or office base), which can lead to the exclusion of workers who may have legitimate claims to be labelled mobile teleworkers.

Part of the reason for taking this broad approach to defining and labelling the type of work examined is to limit the risk of excluding some of the highly heterogeneous jobs and occupations that could be labelled as constituting mobile telework (or whatever the preferred terminology is). Both the ‘mobile’ and ‘tele’ part of the term ‘mobile telework’ are defined to include a wide range of different types of work pattern and practice.

First, the mobile element of the definition encompasses a wide range of patterns in terms of the range and type of locations worked at (Hislop and Axtell 2007). Thus mobile telework covers workers who are totally mobile, and have no static home or office base (such as some types of vehicle roadside recovery engineers) to workers who may have both an office *and* a home base, but who may equally spend a regular amount of time each week travelling to and working at other locations. In addition, mobile telework incorporates workers who travel by a wide range of different means of transport (car, train, plane, etc.) and whose journeys may vary both in terms of their duration, frequency and the extent to which they involve staying away from home overnight (Hardill and Green 2003). For example, the distances travelled by mobile teleworkers can vary enormously. In this volume they range from mobility within a large factory site (see Wiberg, in this volume), to international travel across continents (see Faulconbridge and Beaverstock, in this volume).

Second, the technology or 'tele' element of mobile telework also covers a wide range of technology use patterns. At a minimum level, it incorporates work which may simply involve the regular use of either a mobile phone or a laptop computer (see Axtell and Hislop, in this volume). At the other extreme it incorporates workers who may have a more significant and ongoing need to use a wide range of different mobile computer and communication technologies including the remote accessing of emails (either via BlackBerries or some out-of-office online internet connection), as well as using laptop or handheld computers and mobile phones (see Pica and Sørensen, in this volume).

Mobile telework is thus a label that encompasses a wide range of quite disparate occupations, from mobile office equipment engineers, to management consultants. As will be outlined below, the empirical chapters in the book have been deliberately selected to represent and illustrate this occupational diversity.

Issues

While it is important to take a critical and sceptical perspective on the more extreme and optimistic claims of those advocating the use of m-ICTs, it is equally important not to dismiss the potential significance of the role they can play in facilitating the transformation of work, and social relations in general. Fortunati (2002), reflecting on the implications of mobile phones for society, suggests that they are a potentially revolutionary technology, through the way that they can facilitate new communication patterns, which may produce changes in how space and time are conceptualized and experienced. While space and time are empirically and experientially inseparable, an analytical distinction can be made to consider separately the spatial and temporal implications that m-ICT use, and significant levels of spatial mobility, can have for work and workers.

m-ICTs and the changing spatial architecture of work

The advent of both home teleworking and dispersed/virtual working have had significant implications for the spatial organization of work. First, home teleworking has resulted in work moving into the home, and typically involves workers spending some time working in both their home and at their employer's premises (Halford 2005). Second, the use of virtual working practices where groups of workers who are geographically dispersed collaborate together has resulted in group working becoming increasingly spatially dispersed (Orlikowski 2002). Mobile teleworking, which, as has been illustrated, is becoming more widespread, represents an arguably new and distinctive trajectory in the spatial reorganization of work. With mobile telework, work is increasingly spilling out of private locations such as offices and homes, into public locations such as train carriages, hotel lobbies and cafes. This change, and the significant role that m-ICTs are often argued to play in it, is illustrated by Nunes (2005, p. 137), who suggests that, '[t]he workplace today can be found anywhere electronic networking is possible (planes, trains, hotel rooms, airport lounges…).' An important feature of such work patterns worth noting is that work is not only being carried out in static public locations such as airports, hotel rooms, hotel lobbies and cafes, but also while travelling on work-related journeys in cars (Laurier 2004; Middleton and Cukier 2006), train carriages (Axtell and Hislop 2007; Lyons *et al.* 2007) and aircraft.

The character of many of the types of location that work is creeping into is typically very different from collective office space or a home office, creating distinctive challenges for anyone attempting to work in them. Thus, a growing body of empirical evidence investigating workers' (attempts to) work in such locations questions the 'anytime anyplace' rhetoric by showing how, despite workers being endowed with m-ICTs, the locations they travel through and work in can significantly shape and constrain the realms of what is possible in work terms (Brown and O'Hara 2003; Felstead *et al.* 2005; Perry *et al.* 2001). Halford (2005, p. 20) sums this up by suggesting that, '**where** *work is done makes a difference to working practices and to organisational and personal relationships*', (emphasis in original). Examples of the way in which the spaces occupied by mobile workers constrain what they can do include how the limitations of mobile phone networks make conducting extended phone calls on trains difficult, or how the presence of strangers may inhibit the performance of certain tasks, such as reading confidential documents or making sensitive phone calls.

The constraints on people's efforts to work in such places are many and diverse. First, technological constraints such as whether computers can be plugged into electrical sockets, whether there is reasonable mobile phone network coverage, whether mobile phone use is allowed, or whether wifi internet access is provided affect what, if any, types of m-ICTs can be used. Second, normative constraints shape the type of behaviour regarded as acceptable in different locations (Cooper *et al.* 2002), for example, when loud, extended mobile phone conversations may typically be frowned upon in many public places. Third, the physical/spatial features of locations, which shape the amount of space a person can occupy, can also constrain the type of work tasks which can be carried out, an experience familiar to anyone who has attempted to work while travelling economy on a busy flight. Finally, contextual factors such as the noise levels in different locations can impact on the type of work tasks possible, with tasks requiring concentration being difficult to carry out in noisy public spaces.

However, the relationship between work and space in such locations is not unidirectional, and the features of such spaces are adaptable rather than rigidly fixed. Brown and O'Hara (2003) conclude that the work/space relationship for mobile workers is two-directional and that, not only do the features of particular spaces influence the type of work tasks that can be carried out in them, but that workers have a degree of agency to adapt and modify the character and features of spaces to make them amenable to different types of work tasks. For example, the simple act of reconfiguring the furniture in a room can affect the extent to which it facilitates group working. Further, public spaces such as train carriages can be manipulated to create a reasonable workplace, or a sense of privacy through the colonization of tables, and the strategic use of baggage and seating.

Overall, the spatial implications of mobile teleworking are such that some suggest that the concept of the workplace needs to be fundamentally redefined (Harrison *et al.* 2004).

m-ICTs and the changing temporal rhythms of work

The temporal implications of using m-ICTs in work, and being spatially mobile are potentially equally significant. The 'anytime, anyplace' functionality that mobile phones are argued to offer can change communication patterns through offering what Katz and Aakhus (2002) suggest is the tantalizing potential to have '*perpetual contact*'. This feature of mobile communication devices has the potential to radically change the way people construct and experience time and space through the way they allow people to experience and manage their 'absent presence', their interaction with and participation in a context where they are not physically located (Gergen 2002; Licoppe 2004). Positive aspects of this are that they arguably provide workers with a high level of flexibility to work and communicate at times, and in locations, that are convenient to them (Golden and Geisler 2007). Fortunati (2002) also suggests that a benefit of having and using m-ICTs is that they provide the potential to communicate with intimate and significant others relatively easily, in spite of location and time.

However, there are a number of negative temporal implications to the 'anytime, anyplace' communication potential of m-ICTs. Primarily, they can result in the temporal boundary between work and non-work time becoming blurred and eroded. For example, Prasapolou *et al.*'s (2006) study of Greek professionals talked about how mobile phone use, 'creates a temporal order with increasingly random structural properties which prevents people from drawing definite temporal boundaries around their activities' (p. 280). This conclusion is reinforced by Schlosser's (2002) study of BlackBerry use in the USA, which she found led to the 'blurring of traditional family–work boundaries' (p. 414).

The potential negative implications of this blurring of the work–life boundary are twofold. First, it can result in domestic and non-work matters intruding upon and disrupting work (Townsend and Batchelor 2005). Second, it can have the opposite effect, with work increasingly intruding upon and colonizing non-work time, making it increasingly difficult for people to escape the demands of work. Thus Prasapolou *et al.* (2006) found that taking calls from work colleagues after working hours, which was common, 'allow[ed] work-related activities to penetrate time segments devoted to other spheres of life' (p. 280). Towers *et al.* (2006), who studied how some Canadian office workers made use of home computers and mobile communication technologies felt that this consequence of their use was such that they labelled them '*work-extending technologies*'. Arguably, this helps explain why Hill *et al.* (2003), in a study which compared office-based, home office workers and mobile workers (those who worked from a variety of locations and who were given the technology and autonomy to complete their work tasks in whichever location they deemed most suitable), found that it was the mobile workers who experienced the most negative work–life dynamics.

However, another paradox of m-ICTs is that, while they have the potential to erode people's work–life boundary, they can also serve as tools to actively manage and control this boundary (Golden and Geisler 2007).

Book structure/content

Interest in the topic of spatial mobility and m-ICT use in work spans a number of academic disciplines including business/management, information systems, geography, sociology anthropology and architecture. This book reflects this through the diverse disciplinary backgrounds of the authors. This diversity helps explain the diversity of writing styles, focuses of interest and research methodologies. This heterogeneity is regarded as a positive feature of the volume and it is hoped that the multidisciplinary nature of the contributions will help stimulate a positive dialogue and interaction across disciplines.

The contributions to the books are diverse in two other ways. First, while there is a UK bias among the books' contributors, it also features a significant number of contributions from authors from other parts of the world, including Scandinavia, North America and Australia. Given the broadly global character of the issues examined, it is important to feature empirical and theoretical contributions to represent the ideas and experiences of people from a range of countries.

Second, as outlined, there is a wide diversity in spatial mobility and m-ICT usage patterns in work, with a heterogeneous range of workers being both spatially mobile and using m-ICTs in their work. The contributions to the book also reflect this diversity through the examination and analysis of a range of different types of work from taxi drivers and office equipment service engineers to police officers and professional workers.

The chapters are thematically organized into five parts. First, there is a part on the topic of work space/place, with the contributions in this part being centrally concerned with the spatial implications of workers being mobile and using m-ICTs. The three chapters in this part reveal, not only some of the public places where work is increasingly being performed, and the efforts of people to work in such spaces, but also the impacts and implications such efforts have for all users of such spaces. Holm and Kendall's chapter examines the efforts of professional staff to work in airports using auto-ethnography and observational data, suggesting that such efforts often involve attempting to subvert the logic and experience of being in non-places (Augé 1995). Forlano's chapter by contrast examines the use of mobile phones and wireless internet facilities in some New York cafes and public spaces and finds that working in such spaces can facilitate a form of non-hierarchical, community-based organizing. Finally, Jensen's chapter concludes the section and draws on a diverse range of influences to speculate on the potential negative consequences that may flow from mobile work, and mobile technology usage increasingly colonizing public spaces, such as a reduced sense of community identity and orientation.

Part II shifts focus to examine the topic of work-related travel. Pooley's chapter supplies a valuable historical context in examining the changes in work-related travel patterns in the UK, outlining how the character (such as distance and mode of transport) and experience of such journeys evolved over the course of the twentieth century. Lyons *et al.*'s chapter, based on a multi-method study of travel time usage on UK trains, examines what business travellers do on work-related journeys. Their study challenges the idea of travel time as 'wasted time' by revealing the extent to which the workers studied were able to make such time productive, with such work efforts often involving the use of mobile phones and laptop computers. Faulconbridge and Beaverstock's chapter concludes this section by looking at the role international travel can play in building a sense of common culture and sustaining ongoing social relations within large, geographically dispersed law firms. However, they also conclude that geography matters, as the frequency with which particular locations are visited is shaped by the extent to which they are connected to international travel routes.

The third part in the book contains a diverse number of case studies of mobile work practices. These chapters reveal the heterogeneity that exists in the occupational characteristics of mobile workers, their mobility patterns and the types of issue they face. Axtell and Hislop's chapter opens this part by examining whether having an office base affects the extent to which some UK-based service engineers feel isolated. The fact that those without a permanent work base did not experience significantly greater feelings of isolation is explained by a number of factors such as their use of mobile phones to maintain contact with colleagues, and their efforts to organize informal meetings with colleagues during their work time. Wiberg's chapter looks at spatial mobility and mobile technology use within a large factory complex at a copper process plant in Sweden. Mobile technologies are suggested as having the potential to help solve production problems requiring access to computer data at remote locations. Elaluf-Calderwood and Sørensen, by contrast, examine how the use of mobile technologies by London taxi drivers affects their work. Not only did such technologies increase the level of intercommunication and interdependence between drivers, but they also had the potential to deskill their work. Pica and Sørensen's chapter examines the use of mobile technologies by UK operation police officers. Their analysis challenges the 'anytime, anywhere' rhetoric by suggesting that the context of use is important and affects the extent to which mobile technologies are utilized.

Fourth, the last major part of the book examines the home–work dynamics of mobile workers who use m-ICTs. As outlined, existing research and writing suggests that the blurring of the work–non-work boundary and the intrusion of work into non-work time and space represents one of the largest potential negative consequences of work-related mobile technology use. This, relatively pessimistic scenario is reinforced by most of the chapters in this section. First, Lowry and Moskos utilize existing work–life balance theory to conceptualize the typically negative impact that mobile phone use had for the diverse sample of Australian workers they examined. Townsend and Batchelor present the findings of another Australian study of mobile phone use, this time a survey of some timeshare holiday complex managers. They found that the use of these technologies blurred the work–non-work boundary to such an extent that, not only were some of their respondents most intimate moments occasionally interrupted by work, but that work activities and time were also interrupted by non-work calls from family and friends. The third chapter in this part, by Bergman and Gustafson, develops and applies the concept of 'availability' to understand how the demands that workers undertake work-related journeys in three quite different Swedish organizations affected their availability to participate in family life. Their study found that extensive work-related travel did have negative consequences for people's availability to participate in family activities, but that such demands were most typically experienced by managerial and professional staff, who were typically men. Finally, Middleton's chapter, which draws on an analysis of media articles in North America on BlackBerry usage revealed quite different perceptions by BlackBerry users and their family members of whether and how their use affected their work–life balance. Fundamentally, the behaviours the BlackBerry users felt demonstrated their ability to effectively balance their work and non-work commitments (such as taking it with them to family and social events) were interpreted by their friends and family as being visible signs of a work-life imbalance.

The book then concludes with a final part reflecting on the social policy implications of the growth in workers' spatial mobility and m-ICT usage.

Notes

1 Arnold focuses solely on mobile phones. However, his arguments are relevant to all types of mobile communication technology.
2 Austria, Belgium, Denmark, Finland, France, Germany, Greece, Ireland, Italy, Luxemburg, Netherlands, Portugal, Spain, Sweden, UK.
3 This varied from 20 per cent in Finland, Ireland, Sweden and the Netherlands, to only 7 per cent in Luxemburg and 4 per cent in Portugal.
4 Bergman and Gustafson's chapter examined work-related travel only, and not m-ICT usage.

References

Andriessen, E. and Vartianinen, M. (2006) *Mobile Virtual Work: A New Paradigm?* New York: Springer.

Arnold, M. (2003) 'On the Phenomenology of Echnology: the "Janus-Faces" of Mobile Phones', Information and Organisation, 13: 231–256.

Axtell, C. and Hislop, D. (2007) '"All Aboard!": Trains as Mobile Offices', paper presented at the British Psychological Society, Division of Occupational Psychology Conference, 10–12 January 2007, Bristol.

Brodt, T. and Verburg, R. (2007) 'Managing Mobile Work – Insights from European Practice', *New Technology Work and Employment*, 22, 1: 52–65.

Brown, B. and O'Hara, K. (2003) 'Place as a Practical Concern of Mobile Workers', *Environment and Planning A*, 35: 1565–1587.

Cooper, G., Green, N. Murtagh, G. and Harper, R. (2002) 'Mobile Society? Technology, Distance and Presence', in S. Woolgar (ed.), *Virtual Society? Technology, Cyberbole, Reality*. Oxford: Oxford University Press, pp. 286–301.

Daniels, K., Lamond, D. and Standen, P. (2001) 'Teleworking: Frameworks for Organizational Research', Journal of Management Studies, 38, 8: 1151–1185.

Felstead, A., Jewson, N. and Walters, S. (2005) *Changing Places of Work*. Basingstoke: Palgrave Macmillan.

Fortunati, L. (2002) 'The Mobile Phone: Towards New Categories and Social Relations', *Information, Communication and Society*, 5/4: 513–528.

Gareis, K., Lilischkis, S. and Mentrup, A. (2006) 'Mapping the Mobile eWorkforce in Europe', in E. Andriessen and M. Vartiainen (eds), *Mobile Virtual Work: A New Paradigm?*, New York: Springer.

Gergen, K. (2002) 'The Challenge of Absent Presence', in J. Katz and M. Aakhus (eds), *Perpetual Contact: Mobile Communication, Private Talk, Public Performance*, Cambridge: Cambridge University Press, pp. 227–241.

Golden, A. and Geisler, C. (2007) 'Work–Life Boundary Management and the Personal Digital Assistant', *Human Relations*, 60: 519–551.

Gustafson, P. (2006) 'Work-Related Travel, Gender and Family Obligations', *Work, Employment and Society*, 20: 513–530.

Haddon, L. and Brynin, M. (2005) 'The Character of Telework and the Characteristics of Teleworkers', *New Technology, Work and Employment*, 20, 1: 34–46.

Halford, S. (2005) 'Hybrid Workspace: Re-spatialisation of Work, Organisation and Management', *New Technology, Work and Employment*, 20, 1: 19–33.

Hardill, I. and Green, A. (2003) 'Remote Working: Altering the Spatial Contours of Work and Home in the New Economy', *New Technology, Work and Employment*, 18, 3: 158–165.

Harrison, A., Wheeler, P. and Whitehead, C. (2004) *The Distributed Workplace: Sustainable Work Environments*. London: Spon Press.

Hill, E., Ferris, M. and Martinson, V. (2003) 'Does it Matter Where You Work? A Comparison of How Three Work Venues (Traditional Office, Virtual Office, and Home Office) Influence Aspects of Work and Personal/Family Life', *Journal of Vocational Behaviour*, 62: 220–241.

Hislop, D. and Axtell, C. (2007) 'The Neglect of Spatial Mobility in Contemporary Studies of Work: The Case of Telework', *New Technology, Work and Employment*, 22: 34–51.

Katz, J. and Aakhus, M. (2002) *Perpetual Contact: Mobile Communication, Private Talk, Public Performance*. Cambridge: Cambridge University Press.

Kaufmann, V. (2002) *Rethinking Mobility: Contemporary Sociology*. Aldershot: Ashgate.

Kellerman, A. (2006) *Personal Mobilities*. New York: Routledge.

Larsen, J., Urry, J. and Axhausen, K. (2006) *Mobilities, Networks and Geographies*. Aldershot: Ashgate.

Laurier, E. (2004) 'Doing Office Work on the Motorway', *Theory, Culture and Society*, 21, 4/5: 261–277.

Licoppe, C. (2004) 'Connected' Presence: The Emergence of a New Repertoire for Managing Social Relationships in a Changing Communication Technoscape', *Environment and Planning D: Society and Space*, 22: 135–156.

LSE/Carphone Warehouse (2006) The Mobile Life Report: How Mobile Phones Change the Way We Live. London School of Economics and Carphone Warehouse. Downloaded from: www.mobilelife2006.co.uk.

Lyons, G., Holley, D. and Jain, J. (2007) 'The Use of Travel Time by Rail Passengers in Great Britain', *Transportation Research Part A: Policy and Practice*, 41, 1: 107–120.

Nunes, F. (2005) 'Most Relevant Enablers and Constraints Influencing the Spread of Telework in Portugal', *New Technology, Work and Employment*, 20, 2: 133–149.

Orlikowski, W. (2002) 'Knowing in Practice: Enacting a Collective Capability in Distributed Organizing', *Organization Science*, 13, 3: 249–273.

Perry, M., O'Hara, K., Sellen, A., Brown, B. and Harper, R. (2001) 'Dealing with Mobility: Understanding Access Anytime, Anywhere', *ACM Transactions on Computer–Human Interaction*, 8, 4: 323–347.

Prasapolou, E., Pouloudi, A. and Panteli, N. (2006) 'Enacting New Temporal Boundaries: The Role of Mobile Phones', *European Journal of Information Systems*, 15: 277–284.

Schlosser, F. (2002) 'So, How Do People Really Use Their Handheld Devices? An Interactive Study of Wireless Technology Use', *Journal of Organizational Behaviour*, 23: 401–423.

Sheller, M. and Urry, J. (2006) *Mobile Technologies of the City*. New York: Routledge.

Sullivan, C. (2003) 'What's in a Name? Definitions and Conceptualisations of Teleworking and Homeworking', *New Technology, Work and Employment*, 18, 3: 158–165.

Towers, I., Duxbury, L., Higgins, C. and Thomas, J. (2006) 'Time Thieves and Space Invaders: Technology, Work and Organization', *Journal of Organizational Change Management*, 19, 5: 593–618.

Townsend, K. and Batchelor, L. (2005) 'Managing Mobile Phones: A Work/Non-work Collision in Small Business', *New Technology, Work and Employ*ment, 20, 3: 259–267.

Urry, J. (2000) *Sociology beyond Societies: Mobilities for the Twenty-first Century*. London: Routledge.

Virilio, P. (1999) *Polar Inertia*. London: Sage.

Wajcman, J. (2006) 'New Connections: Social Studies of Science and Technology and Studies of Work', *Work, Employment and Society*, 20, 4: 773–786.

Woolgar, S. (2002) *Virtual Society? Technology, Cyberbole, Reality*. Oxford: Oxford University Press.

Part I

Work space/place

2 Working on the move

Subverting the logic of non-space

John Holm and Gavin Kendall

Introduction: analysing workspace with classical sociology

The classical sociology of work has had some interest in the spaces of work, and less interest in technology as an actor in such workspaces. Marx and Engels, certainly, emphasize the factory as a particular milieu for the development of capitalism (see the famous chapter on 'The Factory' in Marx 1972). One can also discern a similar emphasis in Simmel's (2002) account of how the space of the industrialized city gave rise to a series of psychological comportments crucial to the development of modern life, particularly the blasé attitude and the acceptance of exchange rather than use as the main principle of the economy. It is fair to say that the sociology of work after Marx and Simmel was especially concerned with questions of class relations rather than spatio-cultural or technological questions (although the cultural could often be understood as flowing directly from the economic); so, for example, John Goldthorpe (1963), who did some important work on critiques of embourgeoisement, and Harry Braverman (1974), who developed the theme of the alienating powers of modern scientific work organization, are quite typical of the major concerns of the subdiscipline. It is not so much that technology is absent from these accounts, or that the spaces of work are not acknowledged; it is rather that, as Bruno Latour (1993) puts it, most sociologists are intent on 'purifying' human and non-human actors; human actors are what really matter to sociologists (especially, in the sociology of work, in the form of class or gender relations), and even as non-human actors are introduced into the story, the most that they can hope for is a supporting role.

A second tendency in classical sociological accounts of workspaces – one that owes much to Max Weber – is the realization of the importance of unintended consequences in social development and change. Not only do things go wrong – individuals' social actions do not have their intended effects – but also things can go wrong in quite productive and innovative ways. Individuals' beliefs and actions constantly lead to unanticipated and irrational outcomes – a kind of constant failure of attempts to order the social world – and emergent social structures do not map neatly onto social motives. The *locus classicus* of this position is, of course Weber (2001), which analyses not just how 'rational' capitalism is born out of 'irrational' beliefs and practices, but also how a system of other-worldly beliefs could come to completely reconfigure this-worldly practices (see also Merton 1936). This emphasis on unintended consequences in workspaces can be seen in Habermas's (1989) discussion of the important role of the British coffee houses, the French salons and the German *Tischgesellschaften* in the emergence of a public sphere. These coffee houses and so forth were not only the place for the emergence of a society of letters, but they also promoted a sense of common humanity and equality: "the authority of the better argument could assert itself against that of social hierarchy and in the end carry the day meant, in the thought of the day, the parity of 'common humanity'" (Habermas 1989: 36). These coffee houses were wedded to the intellectual climate of the day not just because they were an informal venue for discussion, but also because they operated as de facto offices and 'drop zones' for the editors of the periodicals (Habermas 1989: 42). A new world of work and a new series of work practices emerge quite contingently. Habermas's work reminds us of the way in which in many modern societies, the world of work has become intimately connected to the use of coffee – to divide work time, to stimulate tiring workers, and to provide a setting in which work discussions can be humanized.

Again, while this emphasis on unintended consequences is incredibly valuable, the deficit in classical sociology is in the refusal to really 'see' the work done by non-human actors. Habermas, for example, is eager to move on to the social relations of the public sphere, and is less interested in dwelling on how the social setting of the coffee house itself, and the form of drop boxes, might shape the possibilities for action. Our chapter seeks to remedy this theoretical and analytical blindspot by suggesting a framework for thinking about the conditions under which work on the move is successful. Drawing on John Law's (2002) account of the multiple spatiality of objects, the argument developed here is that working on the move involves the production of a workplace that is both 'physical' and 'networked'.

Changes to the modern workspace: working on the move

Where people are choosing to work is changing (Harrison *et al.* 2004). Jeremy Myerson and Phillip Ross (2006) provide a recent attempt to delimit systematically the changing workplace. One aspect of this changing workplace, what Myerson and Ross called the 'Agora – the public workplace', is especially significant for our purposes. For Myerson and Ross, the Agora connects 'office life to the wider world of culture, leisure and transport' (2006: 114), and they specifically look at airport business lounges as part of this typology. They argue that airport business lounges have a variety of settings, reflecting the needs of a 'nomadic workforce that has to "work on the pause"' (p. 144). However, while Myerson and Ross tell us much about spaces of work, they do not tell us much about the work practices in those spaces, nor do they tell us much about the practices that make space into workplace.

Felstead *et al.*, who also focus on working on the move (2005: 136–175), make a valuable contribution to understanding how work is done in places of transit. In particular their work is an alternative to what they identify as 'the simplistic message ... that technological artefacts – and in particular, the laptop and mobile phone – are all that is needed to work anywhere/anytime' (p. 136). Instead, they suggest that working on the move is a constant process of taking 'corrective action'. They conclude that successful working on the move depends in a large part on how workers plan to work, noting how tasks are scheduled for particular places or how provisions are made for failure. However, what Felstead *et al.* do not tell us is how these spaces are made into workplaces. Nor do they provide any theoretical purchase on understanding the efficacy of such practices.

The risk that Felstead *et al.* identify, of treating working on the move too simplistically, is pervasive. Even nuanced analyses of the emerging workspaces of the knowledge economy tend to erase the practice of working on the move. What is particularly interesting is how the commentary stops at the design intent of space, without looking closely at the space in use – an error the devoted Weberian would not be likely to make. For example, many texts present business lounges in airports as the archetypical workspace for nomadic workers (see, for example, Myerson and Ross 2006: 144–145). However, airport lounges are not quite as 'coherent' or as facilitative of work as the rhetoric of design implies. Our survey of business lounges in Sydney, Hong Kong, Shanghai, London and Singapore reveal that the provision of workspace is low in comparison to the total capacity of the lounge (approximately one workspace for every fifty people). Nor do these workspaces appear to be the first-choice spaces of work for those in the lounge. Instead most business users opt to sit at 'café'-style tables with their laptops and beverage of choice. Breure and van Meel (2003)'s research into how business travellers use airport lounges supports this observation.

Theories for mobile work

Theorizing how business travellers make transit spaces into workspaces requires a move away from the logic of designed functionality. Rather, we have to focus on how these spaces are *made* to work. As Felstead *et al.* note 'those working on the move have to take corrective action to cope with the deficiencies of public environments as places of work and mould their work activities around what is possible and acceptable in different venues' (p. 136).

The resources we draw on to theorize these spaces include Lévi-Strauss's (1966) notion of the bricoleur and Heidegger's (1978) notion of 'ready-to-hand'. In short, our argument is that the business traveller has to adapt the physical environment to fashion ad hoc solutions to everyday problems, making workspace from any space. Our argument is also aware of Weber's notion of inbuilt failure: social motives are unlikely to be played out successfully, and emergent social structures are likely to be compromises upon, or even contradictions of, intentions and programmes.

We begin this work of theory making by noting that the bricoleur creatively recombines existing resources to produce something that is frequently subversive. Here we mean subversive as simply 'other' to the predominant logics of space:

> There still exists among ourselves an activity which on the technical plane gives us quite a good understanding of what a science we prefer to call 'prior' rather than 'primitive', could have been on the plane of speculation. This is what is commonly called 'bricolage' in French. In its old sense the verb 'bricoler' applied to ball games and billiards, to hunting, shooting and riding. It was however always used with reference to some extraneous movement: a ball rebounding, a dog straying or a horse swerving from its direct course to avoid an obstacle. And in our own time the 'bricoleur' is still someone who works with his hands and uses devious means compared to those of a craftsman. The characteristic feature of mythical thought is that it expresses itself by means of a heterogeneous repertoire which, even if extensive, is nevertheless limited. It has to use this repertoire, however, whatever the task in hand because it has nothing else at its disposal.
>
> (Lévi-Strauss 1966: 19)

Importantly, Lévi-Strauss's work draws attention to the fact that objects cannot be understood in a purely utilitarian way, but are the medium through which people think about their world. The structuralist impulse in Lévi-Strauss's work gives us a sense of the available logics of use and classification within which people can work, but it is always possible to subvert these logics, to use objects for new sorts of thinkings and for inventing a new fluidity of meaning. As he famously says, objects may be chosen by a culture or an individual because they are 'good to think' (*bonnes à penser*) (Lévi-Strauss 1962: 162; see also Woodward 2007: 64–67) – while objects 'afford' (cf. Gibson 1979) certain possibilities, they are not entirely open.

In the airport itself, the logics are logics of flow: the processing of people and objects through space (Augé 1995). This logic does not completely saturate the space of the airport, since airports are rapidly becoming spaces of consumption (Lloyd 2003), and even – although not quite perfectly – spaces of work (Breure and van Meel 2003: 175). We explore how the bricoleur makes space by looking at how a business traveller, denied the luxuries and technical facilities of the business lounge, makes workspace in the waiting areas of the airport. Here we note the production of place that sits uncomfortably within the other spatial logics of the airport, and note the inability of the business traveller to leave his bag unattended.

The bricoleur's activities have an implicit spatiality that corresponds to Heidegger's notion of 'region'. For Heidegger, region was a space determined in relation to our activities (Arisaka 1995: 459):

> The sort of space we deal with in our daily activity is 'functional' or *zuhanden* (ready-to-hand), and Heidegger's term for it is 'region.' The places

> where we work and live – the office, the park, the kitchen, etc. – all have different regions which organize our activities and determine the locations of available 'equipment.' Regions differ from space viewed as a 'container' in that regions are the 'referential' system of our context of activities. This is to say that regions are essentially indexical. The indexical 'here' does not identify a point A in a neutral, container-like space, but rather, our spatial activities determine a 'here' with respect to the things we deal with and the way we move. Regions are inherently organized by activities which emanate from a center of action.
>
> (Arisaka 1995: 458)

These ideas of 'ready-to-hand' and 'region' have two implications. First, there is an implicit physical spatiality of 'ready-to-hand' that is obvious in the activities of the business traveller who makes workplace through the process of drawing together a heterogeneous assemblage of objects. In the waiting area of the airport, he brings together the table, the chair, the laptop, and so on. However, the traveller's activities are not limited to physical space. He also brings into the relation non-material things. The laptop connects, via wireless technology perhaps, to an information network that includes but is not limited to email servers and internet servers. The mobile phone similarly connects ethereal things together for the business user. The second implication, therefore, is that what is ready-to-hand is not limited to physical things. Annmarie Mol and John Law (1994) have developed a notion of region that includes both physical and 'networked' relationalities, and which is useful for our analysis here. More recently, Law (2002) has extended this logic, noting that objects rely on two spatialities for their perpetuation. On the one hand, they require stability in Euclidean space. On the other hand, they require stability in what he called 'network space'. The reason why objects require stability in Euclidean space is obvious: having a laptop 'reconfigured' by it dropping down a flight of stairs makes it more challenging to use as a laptop. Why objects require stability in network space is less obvious.

Network space refers to the arrangements of a heterogeneous materiality required for the functioning of objects. Law exemplified his argument, coincidentally also involving travellers, through a discussion of Portuguese sailing vessels: carracks. He argued that these vessels were, unsurprisingly, required to remain stable in Euclidean space in order to sail. But he noted that the vessels also had to maintain a set of relations within a stable network. At one level this refers to the functioning of the carracks in terms of the practice of sailing. However, it also included maintaining a set of relations in terms of navigation between charts, navigators and the stars (among other things). Finally, the carracks also needed to maintain a network of relations to a Portuguese imperial system that included 'ports, vessels, military dispositions, markets and merchants' (Law 2002: 93).

When we think of the activities of the bricoleur, both types of spatiality are important. In subsequent sections, we want to consider various scenarios that illustrate the importance of considering both types in terms of making workplace in spaces of transit. The first scenario that we consider is the importance of configuring Euclidean space. We also look at behaviours as part of network spatiality and the way that other logics of space impinge on the workplace. The second scenario looks at the configuration of 'virtual' space, arguing that the virtual is a misnomer for the heterogeneous materiality of objects. It is for this second configuration that Law's notion of multiple spatialities has the most relevance.

Spaces of work

What do we mean by mobile working? Laurier (2004) has noted that mobile working requires a significant degree of planning. Specifically his informants made provision for doing different kinds of work in different places. His work, focused on travelling salespeople, highlighted the way that mobile workers allocated tasks to various parts of their journey as appropriate. Telephone calls, for example, were made while driving on the motorways; paperwork was reserved for breaks at services or the hotel at night. Above all, his informants made provision for things going wrong; strategies for coping were developed in advanced, based on experiences in the past. This characteristic of planning for failure is very important for the business traveller.

In part this planning is characterized by what the traveller takes with them (Lyons and Urry 2005). Most travellers carry a laptop, a mobile phone, wireless data cards or wifi. Partly this planning is characterized by other arrangements that the traveller has made, such as remote access to their 'home' network, or placing key documents in online document stores for access later on. But it is also characterized by an expectation of certain facilities in places they are going to experience.

However, these preparations often go wrong. The places that travellers find themselves are frequently not what they expect, and frequently not amenable to working. Places of transit, for example, are rarely designed as workplaces. Airports, train stations and so on are primarily designed to facilitate the flow of passengers (Castells 1996), or perhaps these days the flow of consumption (Lloyd 2003). Augé's commentary on these spaces identifies them as 'non-places', 'space which cannot be defined as relational, or historical, or concerned with identity' (1995: 77). Rather, non-places with their peculiar contractual relations (i.e. the ticket for travel) strip away identities, producing the 'traveller' (p. 103).

In this context we can see the process of creating workspace in this space of flows as a process of re-establishing identity via relations with external networks that subvert the logic of the non-place. This subversion of non-place takes many forms, and it is useful to review four of them here.[1] The first is where the space of flow becomes a space of work.

Using the physical

The most obvious place to begin is the re-establishment, or rather, re-embracing of identity as the traveller becomes the worker. As is alluded to by Augé, the identity of the traveller changes as they enter the business lounge at airports, and it is worth fleshing out this brief description further. Augé notes that the business lounge allows the traveller to reconnect with the world of work, allowing him or her to 'make telephone calls, use a photocopier or Minitel'; in short, to transform from traveller to worker (Augé 1995: 4).

Enter a modern airport business lounge and, while the technology has changed in that 'multi-function devices' have replaced the photocopier, the result is the same. The provision varies from lounge to lounge, ranging from the relatively primitive and bare workstations at Heathrow, to the luxury of Hong Kong's new business lounge that provides Bang and Olufsen phones for those who sit on leather Vitra chairs, working at modern dual-core computers with high-speed internet access.

However, the transformation that takes place is common to all these environments. At these locations, people work. For the time they are there, the hassles of travel diminish into the background and the hassles of work, particularly those delivered via email, come to the fore.

An example of this is a recent business trip one of us made from Mumbai to Brisbane, via Bangkok and Hong Kong. The journey began in Mumbai at 4.50 am, meaning that this business traveller did not check his email before leaving the hotel at 2.15 am. Neither did he bother to check his email at Mumbai international airport, instead choosing to eat breakfast and try to get some more sleep. However, on touching down in Hong Kong, he did use his BlackBerry to see if anything required his attention, or whether he was free to browse the duty free shops in one of the world's best airports for shopping. Unfortunately, it was close to the month's end and the company accountant required some information about the projects he was managing. So he went to the business lounge, grabbed a freshly made latte and went to the business centre. There he printed the WIP (work in progress) report, provided via the company's intranet site, and settled down to itemize the invoices for the month. He used the computers in the business centre both to access and print the WIP.

The choice of business centre was deliberate because he needed to spread the WIP report out to compare last month's completed work against what was invoiced, to capture any outstanding hours from the previous month in the current month's invoices. Additionally, having the WIP also allowed him to update the various project plans to see which projects were on schedule and which ones might have some issues that would require his attention. This work he did on his work laptop as all the project plans were saved there. Some three hours later, the work was completed and the business traveller returned to being a simple tourist who needed to purchase gifts for his family. The duty free shops obligingly took his money.

Using the virtual

However, the contemporary worker does not necessarily need the business lounge to complete his or her tasks. Nowadays most airlines allow their passengers to turn on their mobile devices, phones and BlackBerries, while they deplane. The quiet beeps and tones of missed calls, of new messages and new emails adds to the chaos of the emptying plane. The frantic thumbing of BlackBerry users, scrolling through emails to see what is urgent and what is not, is now characteristic of those waiting to leave. The technological devices have enabled the traveller to reconnect with work: the distinction between those who do and those who do not is an obvious marker of modern status.

Recently one of us, travelling for personal reasons for once, thought that one passenger who had checked his email was going to keel over, as the veins in his neck and temple started throbbing like a character in some B-movie. More seriously, following such passengers off the plane often reveals the extent to which such work is carried out on the pause (to echo Myerson and Ross), in that emails are checked while queuing to leave the plane, subsequent phone calls are made walking along the pier, with protagonists sometimes stopping to check numbers in diaries or Filofaxes. On occasion, people may stop in a vacant gate to email from their laptops if necessary, or they may choose to work in the taxi on their way from the airport. What is common here is that all this work relies on the use of virtual connections to other resources that allow the traveller once more to do work.

But these examples focus too much on simply either the environment or technology. In part this is Felstead *et al.*'s complaint, that too frequently mobile working is simply equated with mobile technology such as BlackBerries, mobile phones and laptops. Law (1986) noted that Portuguese imperialism relied on documents, devices and drilled people. In particular that navigation relied equally on maps, sextants and people who were drilled in the use of these instruments. Understanding the production of the workplace requires understanding how a constellation of heterogeneous elements is established, a process that is frequently fortuitous. The next two sets of examples reveal the process of establishing the constellation, and the frequently subversive (and perhaps sometimes devious?) activities required.

Subverting the physical

First, the subversion of non-place can have a particularly physical aspect. The airport, as a space of flow and consumption, is subverted frequently by travellers doing 'other' to this logic. This can be seen by the backpacker sleeping on the waiting seats, stretched uncomfortably across the seats specifically designed to be uncomfortable if used in this fashion. And perhaps this subversion can be seen even more clearly with those who have unplugged a machine, in this case a commercial massage chair, in order to charge their iPods before once more boarding a flight. Making workspaces in spaces of transit – outside the confines of the proper business spaces of the business lounge – is similarly subversive. One example is the case of a business traveller in a regional airport in Switzerland. Having arrived early for a flight (hoping to get on the earlier departure but not being able to), this individual had four hours to wait. The business lounge was not part of this traveller's loyalty programme, nor was he flying in a sufficient class to gain access, hence it was unavailable to him. Nonetheless, he needed to work during this time and so went about creating a workspace.

The first requirement was for power. His laptop battery was running low, so he found a power point in the general concourse. This power point, probably provided for the cleaners, was located in the base of a pole, and too far from the seats and low tables. So the traveller moved a low table closer (some three feet) and sat on the ground to work. Our traveller created a region in the sense of Heidegger's spatiality of 'ready-to-hand'. He drew together into a direct spatial relationship elements from other logics (the low table is fine to place drinks on but poor for working at, as the business traveller commented: 'I'm too old to sit on the ground like that!').

Additionally, his lack of movement (or consumption) attracted attention, though not significantly enough for him to be prevented from continuing his subversive activities. He received some quizzical looks from security and airport staff, but nobody interfered. Note that this is not always the case, as Coyne *et al.*'s (2005) experiments have shown, where their attempts to hold research seminars in airports saw the security staff move them on.

What is interesting here is that this subversion is only partial. Our business traveller commented that, if he had left his laptop and bag unattended to get a drink or go to the bathroom, as he might have done in the business lounge, he suspected that he might come back to find the bomb squad taking apart his Heideggerian region. This points to a second lesson, that the subversion of this space was temporary. Our business traveller was required to maintain the relationship between the spatial arrangements he made for work, but as soon as he left they must rapidly fall apart. Strum and Latour (1987) made a similar observation about the social order of baboons and its temporary status. In our example, while the business traveller was able to order space materially, the social logic of the airport was still dominant, and threatened to break up the order once it was not held in place by the presence of the traveller. This is suggestive of de Certeau's (1984) analysis of space and of making do.

De Certeau distinguished between 'strategy' and 'tactics' in the configuration of space. 'Strategy' creates a 'proper' space, a space that legitimizes a particular set of power relations. This is similar to Augé's non-place: in these proper spaces, history is erased, leaving only relations (mostly) of capital. Strategy is the airport design that facilitates the flow of passengers and their consumption of duty free items. 'Tactics', on the other hand, do not obey this logic of space. They depend on strategy for possibilities offered (de Certeau 1988: 29), but they are not defined by it. The business traveller depends on the design of the airport for spaces that can become workplaces (e.g. power points for cleaners, etc.) but his activities are not those that the designer intended. Instead, he plays on and with the terrain defined by the designer's strategies. It is his enactment of workplace that is important here. While he sat working on his laptop, plugged into the cleaners' power point, having rearranged the furniture, the space of work was performed. But, as de Certeau noted, he was unable to keep what he won. While he was there, he was able to perform workspace, albeit with some raised eyebrows from security staff.

Interestingly, part of the reason he was left alone was because the behaviours he exhibited were 'normal' in the airport – or, rather, not too abnormal. In part, why these behaviours were tolerated has more to do with the fact that they did not disrupt the other logics of airport space. We can see this in terms of the reluctance the business traveller had to leave his bags to go to the bathroom or to get a drink. Leaving bags like this is (currently) perceived as a security risk. Such a risk causes massive disruption to the everyday working of the airport.

Subverting the virtual

It would be a mistake to view the work of the bricoleur as limited to the physical rearrangement of the furniture. Sometimes the physical is not the problem. There are two examples that we want to look at here. The first has to do with working on a train, when the expected configuration of virtual networks fails, and how the traveller overcomes this failure. This example highlights the importance of the virtual connections that enable work in physical spaces. The second example expands on this theme, highlighting the configuration of devices as enabling networks. The importance of this second example is that it extends Law's (1986) 'documents' to include the settings within devices that enable them.

The first example is somewhat trivial but highlights an important aspect of the mobile workplace. It comes from a traveller in the UK who was making a regular trip from London to Edinburgh on the GNER Mallard train. The Mallards were one of the first trains to have wireless connectivity, enabling travellers to connect to the internet. This business traveller, having made this trip many times, had come to rely on this connectivity. However, on one occasion the wireless was not working. Unfortunately the business traveller had promised a colleague a document by close of business that day, and the train would not make Edinburgh in time to keep this promise.

The traveller, however, was able to transfer the file in question to his mobile phone, via a USB cable. From his phone, he was able to email the document to his colleague. This example is important in that it highlights the 'ready-to-hand' aspects of making workplace. The phone was ready-to-hand. But it was not an obvious choice. The file was of a significant size, almost one megabyte. The reason the document was so large was that it was highly configured, involving some advanced features in Microsoft Word, features which also meant that the document was not supported by the applications on the phone. While it was a Word document, the phone only supported a simplified version of Word. In short, the phone was functioning as email client only. The reason it was not an obvious choice is that the phone network is not a great way to transfer such large files. In this sense the use of the phone is devious, exploiting the technology in a way that, while possible, was not quite as intended. The lesson here is that the technology, ready-to-hand in the case of the mobile, allowed the business traveller to make a connection with a technological network (the email system) that also established/maintained the network of work in terms of exchanging information with a colleague.

Our next example comes from an Australian business traveller who needed to complete a report for a client. Unfortunately, he ran out of time before he had to travel. Fortunately the 'intelligent generosity of his firm', to repeat Augé's phase (1995: 3), had led to his travelling business class. This allowed him to plug into the aircraft's power supply, within the wider seat, and complete his report while travelling from Sydney to Singapore. The 'personal and constant attentive service' (Augé 1995: 4) on the plane meant that he lacked for nothing for the seven hours or so he was in the air.

Having arrived in Singapore, he went to the business class lounge, grabbed a glass of wine, and sat down to email his report to the client. Again the intelligent generosity of the firm provided him with a 3G wireless broadband card which, due to international partnering arrangements on the part of the suppliers, would work not only in Australia but also in a number of other countries, including Singapore. Unfortunately, due to an oversight by his company's operations manager, this particular wireless card was not enabled with international roaming. Given that it was now 11 pm in Australia, this oversight was not immediately rectifiable. The business traveller, however, had other avenues. The business club lounge connected to his frequent flyer membership provided free wireless internet to its users. So he went to the reception desk, leaving his laptop open on the low table where he had sat down to work, to get the login details. Returning to his temporary workplace, he logged into the wireless internet and tried to send his email.

However, the email software, Microsoft Outlook in this instance, was configured for the SMTP server of the provider of the 3G wireless broadband network. This meant that the business traveller's email would not leave his outbox. Frustration was beginning to creep into his typing at this stage; keys were being struck with some force.

Not conceding defeat, the business user called his company's global information technology and communication department's help desk. The technician on duty, based in Maine in the USA, was able to supply a new IP address for another SMTP server that would allow the business traveller to send his email. Eventually the email was sent, and the business traveller relaxed in the lounge until his flight departed.

The interesting aspect of this example is that, physically, not much changed. The changes the business traveller had to effect to make his workplace were all to do with creating a particular set of relations that enabled the space as workspace. These changes, while practically in the realm of IP configurations and the arrangement of technological networks, were also social. Or better, in ANT terms, these provide an example of heterogeneous engineering (Law 1986) where 'bits and pieces from the social, the technical, the conceptual and the textual are fitted together' (Law 2003: 2). The important aspect to the business traveller's engineering is the way he reached out to configure his temporary workspace, drawing on the local IT infrastructure of the wireless in the lounge in Singapore, connecting to email servers in Australia, but reconfiguring his email protocols with the help of technicians in the USA. Working on the pause is not as easy as it first appears.

Conclusion

The literature on mobile working tends to focus on the provision of technology or presumes the realization of the design intent of spaces intended for those working 'on the pause'. What this chapter has focused on is that there are many prerequisites for successful mobile working that include the configuration of the 'ready-to-hand' into a region of work and the successful establishment of network spaces that enable the worker to participate in the kind of global asynchronous working characteristic of mobile working. The importance of these two correctives is that we return to Law's argument that successful expansion beyond an immediate occupied physical requires the assemblage of a heterogeneous materiality of devices, documents and drilled people.

This sort of analysis is, we think, a valuable addition to classical sociological attempts to understand the world of work. The Weberian emphasis on unexpected outcomes still proves a useful basis for any contemporary analysis – the gap between social intentions, beliefs and motives and social structures is still likely to be as wide today as it was seen to be in Weber's time, and it is clear that the modern worker-on-the-move is constantly reacting to the failure of the social and technological systems that surround him or her, constantly reworking them to generate one-off solutions that enable the work network to be repaired.

Note

1 Our methodology here is a combination of ethnography and auto-ethnography, and drawn from a variety of empirical sites. One important source is the direct experiences of being a business traveller: John Holm works as an architectural consultant and researcher with Woods Bagot, an international architecture firm, in which role he has travelled extensively – sometimes he thinks too extensively – in Europe, Asia and Australia. We also observed and spoke with fellow travellers, and traded 'war stories' of working on the pause. In other instances, travelling on planes with other business travellers allowed for informal observational study.

References

Arisaka, Y. (1995) 'On Heidegger's Theory of Space: A Critique of Dreyfus', *Inquiry*, 38, 4: 455–467.

Augé, M. (1995) *Non-places: Introduction to an Anthropology of Supermodernity*. London: Verso.

Braverman, H. (1974) *Labor and Monopoly Capital: The Degradation of Work in the Twentieth Century*. New York: Monthly Review Press.

Breure, A. and van Meel, J. (2003) 'Airport Offices: Facilitating Nomadic Workers', *Facilities*, 21, 7/8: 175–179.

Castells, M. (1996) *The Rise of the Network Society, The Information Age: Economy, Society and Culture*, vol. I. Cambridge, MA and Oxford: Blackwell.

Coyne, R., McMell, D. and Parker, M. (2005) 'Places to Think with Non-Place and Situated Mobile Working', Working Paper Architecture. School of Arts, Culture and Environment, University of Edinburgh.

De Certeau, M. (1984) *The Practice of Everyday Life*. Berkeley: University of California Press.

Felstead, A., Jewson, N. and Walters, S. (2005) *Changing Places of Work*. Basingstoke: Palgrave.

Gibson, J.J. (1979) *The Ecological Approach to Visual Perception*. Boston, MA: Houghton Mifflin.

Goldthorpe, J. (1963) *The Affluent Worker: Political Attitudes and Behaviour*. Cambridge: Cambridge University Press.

Habermas, J. (1989) *The Structural Transformation of the Public Sphere: An Inquiry into a Category of Bourgeois Society*. Oxford: Polity Press.

Harrison, A., Wheeler, P. and Whitehead, C. (2003) *The Distributed Workplace: Sustainable Work Environments*. London: Spon.

Heidegger, M. (1978) *Being and Time*. Oxford: Blackwell.

Latour, B. (1993) *We Have Never Been Modern*. Cambridge, MA: Harvard University Press.

Laurier, E. (2004) 'Doing Office Work on the Motorway', *Theory, Culture & Society*, 21, 4/5: 261–277.

Law, J. (1986) 'On the Methods of Long-Distance Control: Vessels, Navigation and the Portuguese Route to India', in J. Law (ed.), *Power, Action and Belief: A New Sociology of Knowledge?* London, Boston, MA and Henley: Routledge and Kegan Paul, pp. 234–263.

Law, J. (2002) 'Objects and Spaces', *Theory, Culture & Society*, 19, 5/6: 91–105.

Law, J. (2003) 'Notes on the Theory of the Actor Network: Ordering, Strategy and Heterogeneity', published by the Centre for Science Studies, Lancaster University, Lancaster LA1 4YN, at www.comp.lancs.ac.uk/sociology/papers/Law-Notes-on-ANT.pdf.

Lévi-Strauss, C. (1962) *Totemism*. Harmondsworth: Penguin.

Lévi-Strauss, C. (1966) *The Savage Mind*. Chicago, IL: University of Chicago Press.

Lloyd, J. (2003) 'Airport Technology, Travel, and Consumption', *Space and Culture: International Journal of Social Spaces*, 6, 2: 93–109.

Lyons, G. and Urry, J. (2005) 'Travel Time Use in the Information Age', *Transportation Research Part A*, 39: 257–276.

Marx, K. (1972) *Capital*, vol. 1. London: Dent.

Merton, R.K. (1936) 'The Unanticipated Consequences of Purposive Social Action', *American Sociological Review*, 1, 6: 894–904.

Mol, A. and Law, J. (1994) 'Regions, Networks and Fluids: Anaemia and Social Topology', *Social Studies of Science*, 24: 641–671.

Myerson, J. and Ross, P. (2006) *Space to Work: New Office Design*. London: Laurence King.

Simmel, G. (2002) 'The Metropolis and Mental Life', in G. Bridge and S. Watson (eds), *The Blackwell City Reader*. Oxford: Blackwell, pp. 1–19.

Strum, S. and Latour, B. (1987) 'The Meanings of Social: From Baboons to Humans', *Social Science Information*, 26: 783–802.

Weber, M. (2001) *The Protestant Ethic and the Spirit of Capitalism*. London: Routledge.

Woodward, I. (2007) *Understanding Material Culture*. London: Sage.

3 Working on the move

The social and digital ecologies of mobile work places

Laura Forlano

Introduction

This chapter describes the social and digital ecologies of mobile work places. Social and digital ecologies describe human and technological relationships that exist in a particular place. For example, human-to-human, human-to-computer and computer-to-computer relationships might be taken into account as being part of a network of people and technological artifacts. Mobile work places are non-traditional work settings including cafes, parks, airport lounges and other public and semi-public places. In recent years, such locations have become important work places for mobile professionals, in particular, among remote workers, telecommuters, self-employed and freelance workers, and entrepreneurs. The declining cost and widespread use of laptop computers, mobile phones and personal digital assistants (PDAs) as well as the increased deployment of wireless fidelity (WiFi) hotspots contributes to the usefulness of mobile work places. Drawing on a year-long ethnographic study of mobile work places in New York, I will argue that these locations are intense sites of informal interaction, social support, collaboration and community. In addition, I will show how mobile work places blur, and often reverse or contradict, traditional dichotomies such as work and play, online and offline, public and private, presence and co-presence, individual and community, and local and global. In contrast to media representations of mobile work that focus on freedom, convenience and 'anytime, anywhere' use of mobile and wireless technologies, I will illustrate the ways in which mobile work places support emergent and rapidly transforming occupations. I will explain the public nature of mobile work and the deliberate reasons that mobile professionals choose mobile work places. Finally, I will conclude by arguing that mobile work places are examples of the (re-)emergence of a community form of organizing that coexists with hierarchical, market and network forms of organizing.

Theoretical framework

This paper employs the ritual view of communication (Carey, 1988), the social construction of technology (Bijker *et al.*, 1987) and actor-network theory

(Latour, 2005), concepts from communications, and science and technology studies, as the key theoretical frameworks. Carey argues that most American studies of communication employ a 'transmission or transportation view of communication' and the 'effects' tradition that views communication 'basically as a process of transmitting messages at a distance for the purpose of control' (1988). In the last decade, since the mainstream adoption of the Internet, there has been an overwhelming emphasis on the ways in which communications transcends geographic constraints. Carey writes that such studies focus on 'persuasion; attitude change; behavior modification; socialization through the transmission of information, influence or conditioning' (1988).

In contrast to the 'transmission view,' Carey advances a 'ritual view,' which builds on earlier studies of communication by Harold Innis as well as concepts of culture advanced by Clifford Geertz, Raymond Williams and Stuart Hall. Innis (1951) theorized that all media could be identified as either *time-biased* or *space-biased*. Time-biased media such as oral tradition assert their control over the maintenance and preservation of ideas in time while space-biased media such as paper expand the reach of ideas in space for the purposes of control. For the purposes of this study, wireless networks can be understood both as space-biased and as time-biased media. This is because, while wireless networks allow users to connect to the Internet, they are also located in bounded physical and digital spaces where users often commune together.

Carey's 'ritual view' elaborates on Innis's theorizing about time-biased media, asserting the following: first, 'communications is first of all a set of practices, conventions, and forms;' second, 'communication is a process through which shared culture is created, modified, and transformed;' and, third, communication should be 'directed not toward the extension of messages in space but in the maintenance of society in time,' and on the 'sacred ceremony that draws persons together in fellowship and commonality' (1988). By adopting the 'ritual view' as the key theoretical framework, this study seeks to understand the practices and cultures of community wireless organizations and users of WiFi hotspots and the way in which they maintain associations in time.

The social construction of technology understands technologies as being the products of the interplay between historical, economic, political, cultural and social factors. Actor-network theory understands human and technological agents being part of a network in which they have equal status. Actor-network theory is particularly well suited for a study of mobile and wireless networks especially in light of future ubiquitous computing scenarios, which imagine a world of networked people and objects (Weiser, 1991). Another advantage of using actor-network theory as a framework is its emphasis on 'following the user' in order to uncover relevant practices, technologies and places as research sites.

Since the Industrial Revolution, hierarchical and bureaucratic forms of organizing have dominated until very recently. In the current period, since the development and mainstream adoption of the Internet, much emphasis has been placed on network and virtual forms of organizing. These forms emphasize fluidity and adaptability, decentralization, a renegotiation of power relationships and the need to minimize constraints while maximizing possibilities. Network forms are also referred to as 'epistemic communities' in political science, 'communities of practice' in sociology, and 'knowledge networks' in management (Howard, 2002). The key characteristics of network forms are: the use of information technology to integrate across organizational functions and to coordinate geographically dispersed activities; flexible, modular organizational structures that can be adapted; team-based, flat hierarchies and horizontal coordination; and the use of intra- and interorganizational markets (Poole, 1999). Similarly, virtual forms of organizing are geographically distributed; electronically linked; functionally or culturally diverse; and laterally connected, which makes possible highly dynamic processes, contractual relationships, edgeless, permeable boundaries and reconfigurable structures (DeSanctis and Monge, 1999). The term 'virtual' has been widely used to describe new forms of organizing work, which differ from traditional forms in the location of the workers, where and how the work is accomplished; and the basis for relationships between workers and organizations and between organizations (Watson-Manheim *et al.*, 2002). Recent scholarship on the open source movement, music file-sharing, and WiFi-sharing (Benkler, 2006; Noam, 2005; O'Mahony, 2002) has documented the emergence of economic models based on peer production and community forms of organizing and sharing. These models coexist and compete with hierarchical, market and networked forms of organizing.

While there is a rapidly growing body of research on the way in which mobile phones are used (Ito *et al.*, 2005; Katz and Aakhus, 2002; Pedersen and Ling, 2005) and some research on early Internet cafes and cybercafes, this research has not addressed the adoption of WiFi hotspots as mobile work places. These studies have included analyses of user behavior – in particular, videogame behavior – at cafes in Toronto (Middleton, 2003; Powell, 2003); the embedding of local and global culture at cafes in London (Wakeford, 2003); cafes as innovative sites of access to information and communication technology in the UK (Liff and Laegran, 2003; Liff and Steward, 2003); the significance of place for mobile work (Brown and O'Hara, 2003); the relationship between the cybercafe and the community in Scotland (Stewart, 1999); and domestic and public uses of technology at cafes in the UK (Lee, 1999). Few studies have addressed the role of mobile and wireless technologies in (re-)organizing work practices (Gupta, 2004; Mazmanian *et al.*, 2006; Rheingold, 2003). This chapter aims to begin filling this gap.

Methodological framework

This paper draws on a mixed methodology (Axinn and Pearce, 2006; Creswell, 2003; Norman, 1990; Tashakkori and Teddlie, 1998), which combines ethnography, participant observation and in-depth qualitative interviews. In May 2006, I spent over 20 hours observing the wireless Internet users at a popular cafe on the Lower East Side of New York between the hours of 11 am and 9 pm. The cafe was selected because it was popular with a wide variety of freelancers, students and artists in the neighborhood, where it has been located for over ten years. The activities of each of the clientele in the cafe was noted, with specific attention paid to their use of mobile and wireless technologies and their interactions with others in the cafe. Participant observation was also conducted in that, on several occasions, I attempted to work from the cafe myself (unfortunately, without much success). The cafe is located in the heart of New York's historic Jewish neighborhood, the Lower East Side, which is bounded by Houston Street, Canal Street and the Franklin Delano Roosevelt Drive on the East River.[1] This initial field study allowed me to identify patterns of use in order to inform the in-depth interview questions.

Following the month-long observation at the cafe, I developed a survey on the use of wireless networks in cafes, parks and other public spaces. The 40-question online survey was conducted between October 2006 and April 2007, in New York, Montreal and Budapest. The survey was conducted with a small grant from Microsoft Research in partnership with local community wireless organizations: NYCwireless (New York), Île Sans Fil (Montreal) and the Hungarian Wireless Community (Budapest). In New York, the surveys were publicized through fliers, on listservers, via email announcements, and via the login or 'splash' pages of the wireless networks of partner organizations. In New York, the Downtown Alliance, a Lower Manhattan business improvement district, placed a link to the survey on their website. The survey was included in New York City Council Member Gale Brewer's monthly email announcement. In Montreal and Budapest, the survey was only publicized online. The survey was conducted using SurveyMonkey,[2] an online survey tool. The survey resulted in 1362 responses: New York (614), Montreal (370) and Budapest (378). While discussion of the survey results is not within the scope of this paper, the survey provided a valuable way of identifying informants for in-depth interviews.

Following the survey, I conducted in-depth qualitative interviews centered around three sites in New York: Bryant Park, Starbucks and the JetBlue Terminal at JFK. These three sites were chosen because they represent three different types of settings where WiFi hotspots are often deployed: cafes, parks and public spaces, and airport lounges. Following is a short description of the three research sites. First, Bryant Park is a privately managed public park in midtown Manhattan, which is located on Forty-Second Street between Fifth and Sixth Avenues directly behind the New York Public Library. The Bryant Park wireless network was built by NYCwireless, a community wireless organization, in partnership with the Bryant Park Restoration Corporation in 2002. The wireless network at Bryant Park is free to use. Second, Starbucks, an international coffee retailer, has 153 locations in the New York area (within a five-mile radius) where a T-Mobile HotSpot is available.[3] The T-Mobile HotSpot requires customers to pay daily, monthly or annual membership fees in order to access the wireless network. Interviews were conducted in numerous Starbucks locations in Manhattan and Brooklyn. Third, JetBlue is a low-cost airline in the United States. While no interviews were conducted in the JetBlue Terminal at JFK, I asked informants whom I interviewed in other locations about their experiences with the JetBlue wireless network. In addition, on several occasions, I conducted observations at the JetBlue Terminal at JFK. All three case studies were documented through a combination of ethnographic observation, in-depth interviews and photography.

About 80 people – those who reported using the wireless network for two or more hours a day at a cafe, park or other public space – were selected for interview from the New York survey. About 50 agreed to be interviewed. A total of 29 in-depth interviews were conducted with users of WiFi hotspots between October 2006 and April 2007. The interviews with users of WiFi hotspots were one-hour, open-ended interviews, which were conducted at the locations in which the informant reported that they used the wireless Internet most frequently whenever possible. All interviews were documented with notes and recorded digitally. On several occasions, interviews were conducted by phone and recorded via Skype. The interviews were not transcribed.

Of the 29 interviews with users of WiFi hotspots, the following breakdown emerged for the three types of research sites: cafes (20), parks and public spaces (8), and airport lounges (1). It is important to note that individuals had often frequented WiFi hotspots in a number of different locations including cafes, parks and airport lounges so the interviews often reflect their experiences at a number of sites. This paper draws primarily on the interviews conducted in cafes. Among those interviewed, 24 were men and five were women.

Interviews focused on informants who reported in response to the survey that they were full-time or part-time employees, self-employed or entrepreneurs. Fourteen of the informants are full-time employees; 13 are freelance, self-employed or entrepreneurs; and two were unemployed during the time that they reported using WiFi hotspots. Among full-time employees, one works remotely in finance for a DC-based firm and another works remotely in technology sales for a Silicon Valley-based firm. Among freelancers, one works remotely in public relations for a Boston-based firm. Informants worked in a range of occupations. One is a university professor, one is a photo-equipment repairman, one works in finance, one is a graphic illustrator, one works in hospitality, one is a lawyer, one works in media production, four work at non-profit organizations, one is a performer, two work in public relations, seven work in technology, one works in translation, two are Web designers and three are writers. One of the unemployed informants is a homeless blogger. In the descriptions that follow, all informants are identified by pseudonyms in order to protect their privacy.

Mobile work places

The following section draws on a number of theoretical concepts that have been employed to describe the nature of place, in order to develop a definition of mobile work places. Using Suchman's concept of situated action, which describes 'actions taken in the context of particular, concrete circumstances' (1987), the activities of mobile professionals can be analyzed with respect to the presence (or absence) of physical, technological and social factors. On the one hand, mobile work places are the 'third places' that Oldenburg believes are necessary for the functioning of urban social life, which he fears are rapidly disappearing (1989). However, on the other hand, mobile work places are distinct from third places because, rather than being a comfortable and casual place away from home or work, in practice *they are increasingly used as work places*. Castells has articulated the tension between the 'space of flows' – global networks of technology flows – and the 'space of places' – the urban spaces of everyday life (1996). Mobile professionals are simultaneously participating in the 'space of flows' by virtue of their wireless connections to telecommunications and the Internet while, at the same time, cultivating the 'space of places' by forming indisputably local social networks as part of their everyday working life.

The concept of innovation spaces captures the recent interest of firms in designing physical environments that foster innovation and creativity (Moultrie *et al.*, 2007). For example, Motorola invented the successful Razr phone by spinning off a separate project team, locating them in a chic office in downtown Chicago (rather than at the corporate headquarters in the suburbs) and creating an environment where employees from different business units including marketing and engineering could interact closely. As the above narrative illustrates, mobile work places may be seen as innovation spaces for those who use them to stimulate their own productivity, expand their social networks, participate in site-specific work communities and collaborate on projects.

According to Stewart Brand, co-founder of the Global Business Network and founder of the Whole Earth Catalog, 'the most productive people he knows have developed ways to work outside offices, not in them' (Schlosser, 2006). The sonic environment of a mobile work place is strikingly different from the stereotype of a typical white-collar office environment. For example, mobile professionals working in cafes are often surrounded by the loud screeching of the espresso machine as milk is foamed to perfection. While this could be seen as an inconvenience, many mobile workers report that sound is an important stimulant for their work. Those who are distracted or bothered by the ambient noises often use personal music devices such as iPods in order to block out the sounds.

For example, one informant, Adam, a middle-aged freelance writer from Brooklyn, elaborated that, while he rented an office space, he didn't like to work there. 'It's too quiet,' he said. Instead, he explained, 'Starbucks IS my office.' He has been working at a Starbucks for the past five years because he likes to be in a place where he can see people passing by engaged in their daily activities. 'The challenge for office planners is to create flexible, stimulating spaces that are an attractive destination for employees who can choose when and where they work' (Bloemink *et al.*, 2006). According to the design team at Herman Miller, 'The kind of anonymity found in plain sight at Starbucks, the kind of variable stimulation found in libraries and public plazas – these are the new qualities to be fought for in work environments' (Bloemink *et al.*, 2006).

Another indication of the importance of sound is a CD, 'Thriving Office,' which is being marketed to telecommuters and small businesses. The CD boasts two 39-minute tracks – 'Busy' and 'Very Busy' – that replicate 'the sounds people expect to hear from an established company' such as voices, phones and computers. The CD has received accolades from the *Wall Street Journal, Business Week* and National Public Radio as well as management gurus such as Tom Peters. Steelcase's WorkSpace Future Team said, 'This background buzz keeps their energy up and mind in the game,' and the Herman Miller Corporation claimed, 'Research has documented productivity gains of 38%, job satisfaction increases of 175% and stress reduction of 27%' ('Thriving Office,' 2007). Similarly, at mobile work places, people are *simulating the office environment* that they lack as a remote worker, telecommuter, freelancer or self-employed worker.

At the Lower East Side cafe that I studied, the soundscape was a *mélange* of Tom Waits being played over the cafe's speaker system and reggaeton[4] bleeding in from the street through the open door. While the area was once the home of the world's largest Jewish community,[5] it is currently populated with designer boutiques stashed amid the Spanish bodegas and bargain stores during the day. But, in the evening, it becomes one of New York's trendiest spots for nightlife including high-end restaurants, bars and music venues. The cafe – known as a 'Boho café by day, low-key bar by night'[6] – is described as having a 'café society.'

The cafe offers free wireless Internet and, during peak hours (approximately from noon until 6pm), it is often difficult to find a table, which is testament to its popularity as a mobile work place. Overall, the clientele are diverse in terms of race, gender and age. The cafe's physical space is divided into several sections: a cafe, and a bar. The cafe has approximately 11 tables – six small round tables and one small square table that accommodate two people each, four large rectangular tables that accommodate two to three people each – where people can plug in their laptops.[7]

The following section illustrates the way in which mobile work is tied to the transformation of the economy as a whole, the nature of public places themselves and *specific times and places* that are meaningful to individuals. As such, mobile work places are intense sites of informal interaction, social support, collaboration and community.

Working for the algorithm

Mobile work places support emergent occupations, practices and organizational structures that are evolving as the economy as a whole is transformed in the era of globalization and networked computing. Over the past ten years, the number of freelancers – independent contractors, self-employed and temporary workers, entrepreneurs – has increased and is now expected to make up 10–30 percent of the workforce in the United States. In New York, the media and technology industries, including advertising, publishing, film and television, technology and the arts, employ the large majority of freelancers. In addition, in general, freelancers are well educated, earn over 20 percent more than the city's overall median income, and are entrepreneurial, creative and independent workers (Horowitz *et al.*, 2005).

For example, Daniel, a freelance and comedy writer from Boston, is currently employed as a search engine optimizer (SEO). SEOs are an example of a new occupation, which did not exist prior to the widespread adoption of the Internet and, more specifically, to the increased importance of Google and other search engines to navigate the Internet. Like many emergent occupations, the job of an SEO can easily be done on a project basis and requires little face-to-face interaction with companies or clients. The task of an SEO is to create unique articles for websites that have not yet launched in order to increase the ranking of the site on search engines such as Google. SEOs are given a list of keywords that must be used repeatedly in the articles. While the task of writing such articles sounds almost routine, it cannot be done by an automatic program because Google 'knows' the difference between original text and that which is, for example, copied from another website. The Google algorithm discriminates against sites that are merely copied and demotes them to sub-par status, known as 'grey-listing.' Instead, a writer must create unique text for the website despite that fact that it will never be read by anyone at all. As a result, the sites contain a mixture of fact and fiction, research and imagination. Thus, perhaps it should not be surprising that people working in such occupations seek out informal interaction, social support and community at mobile work places.

Making work public

Mobile work places are public or semi-public places. As such, they blur, and often reverse or contradict, traditional dichotomies such as employee and employer, work and play, online and offline, public and private, presence and co-presence, individual and community, and local and global. For example, since a significant number of mobile professionals are freelancers, self-employed workers and entrepreneurs, the distinction between employee and employer is not well defined.

Work and play are blurred through the heterogeneity of activities occurring simultaneously in mobile work places. For example, at the Lower East Side cafe that I studied, while many of the clientele were working on their laptops, others were talking with friends, making mobile telephone calls, eating, playing video games, drinking beer, reading or writing in their journals. In addition, the clientele often spent more than two hours in the cafe both working and socializing intermittently, and sharing beers with other patrons at the end of the work day. Those clientele that knew each other, the 'regulars,' often visited each other's tables throughout the day to take short breaks from their work.

Mobile work places are sites in which online and offline activities coexist. This includes the coexistence of knowledge-work, service-work and unemployment. Many mobile professionals work from this Lower East Side cafe in order to use the free wireless network. However, James, an academic, reported that he specifically comes to the cafe in order to be offline and do his writing. James has access to the Internet at home where he is constantly bombarded with telephone calls and emails. Thus, for him, the cafe represents a haven where, while others are online around him, he can escape from the demands of communication. In addition, while many patrons enter armed with high-end laptops, iPods, PDAs and mobile phones, others do not have access to these technologies. In particular, the cafe is home to a number of drifters including a number of alcoholics and a mentally ill woman who busses tables in exchange for free coffee. The woman, who lives upstairs from the cafe, is assumed to be unemployed and spends long hours in the cafe passing the time. When patrons leave newspapers or coffee mugs on the table, she quickly folds the papers and returns them to a communal rack and clears away the dirty dishes.

Mobile work places also blur the boundaries between private and public in unique and interesting ways. While cafes are clearly private spaces in some ways, when compared to public parks for example, they are generally open to all and attract a wide range of patrons. Mobile workers often use technologies in order to signal their availability for interaction or conversation. For example, laptop screens often serve to indicate when someone is engaged in their work or open to being interrupted. Similarly, iPods and other portable music players are used to create bubbles of privacy in the midst of the public space of the cafe. The practice of donning headphones in order to shield oneself from the ambient café noise the music being broadcast over the loudspeakers is widespread among mobile professionals. In addition, while mobile phone calls were sometimes conducted in the cafe itself, it is far more common for patrons to leave the cafe and pace up and down the street while making such calls. This is interesting in that it reverses the commonly held notion that inside space is private space while outside space is public space. In this case, mobile workers leave the public space of the cafe in search of a more private space on the street where they can carry on their personal or business conversations.

Finally, mobile work places are sites of temporal, spatial and project flexibility. While work in a traditional white-collar office environment typically begins at 9 am, the hours of a mobile work place are not dictated by economic forces alone but rather by a mixture of social, cultural and personal norms. For example, this Lower East Side cafe was rarely crowded before noon when the first laptop-toting clientele would typically arrive. Before noon, the clientele mostly consisted of those who came to have coffee and read the newspaper in a relaxed environment. Similarly, after 8 pm, the lights dim, the music gets louder and the cafe is rendered into a bar environment. Mobile work places offer the possibility of spatial flexibility. Not only is it common for mobile workers to move from place to place throughout the course of their day, the cafe's wireless network also allows them to move from table to table without sacrificing their connection to the Internet. For example, one regular mentioned that he would often work from one cafe in the afternoons and then move to another in the evenings. This was partly for a change of scenery and partly due to the fact that the first cafe became a bar in the evenings. In addition, during peak hours, people often began working at a table that was less desirable for its small size, cramped location or lack of proximity to electrical outlets. Thus, when a more desirable table opened up, it was common for that person to move to a more optimum location. Interestingly, people often seemed to prefer to share one of the larger, rectangular tables rather than working alone at the small, round tables. This is partly because the large, rectangular tables offered more space and were slightly more conveniently located regarding plugging into electric outlets. Thus, rather than remaining in one location for the duration of the work day, it was common for clientele to table-hop until their ideal table was reached. Finally, it is assumed that mobile workers have some degree of control over the type of projects that they are working on and activities that they are participating in at any given time and place. Thus, it can be said that mobile work places offer possibilities for project flexibility in addition to their characteristic temporal and spatial flexibility.

Collaboration and community

Casual conversations and informal interactions, often referred to as 'water-cooler' conversations are known to build trust, create social support and promote innovation and collaboration in traditional office settings. Since mobile professionals are often not physically present at a traditional office, research has focused on the impact of electronic communication on informal interaction, finding that organizational communication typically declines as the use of email increases (Sarbaugh-Thompson and Feldman, 1998). However, interestingly with respect to the case of mobile work places, ample opportunities for informal interaction exist. This is because, due to the relative flexibility of mobile work places, each interaction represents a negotiation for location, electricity, connectivity and security.

Laptop users needing to plug in must often negotiate their way to a slot at the nearest power outlet, or, when seeking to connect, they might ask a neighbor the name of the wireless network. For example, at Bryant Park in October 2006, I overheard one man with a laptop say to another, 'What is the network SSID?' Finally, for those who spend long hours working at mobile work places, the security of their equipment and belongings is vital especially since they don't want to lose their valuable seat whenever they need to make a phone call, eat lunch or go to the bathroom. In order to maintain their location, while being granted some flexibility, they may turn to the nearest person to determine whether they are trustworthy. They may initiate the interaction by making a comment about the music and waiting to see how their neighbor answers. Based on their neighbor's reaction, they may decide whether or not to leave their laptop in the care of the stranger. In my observation at the Lower East Side cafe, to my surprise, people often left their laptops completely unattended for long periods of time – without asking a stranger to monitor it. In this example, the community plays a surveillance function in order to maintain a casual and relaxed but secure environment that makes others feel comfortable leaving their belongings unattended. However, one Starbucks in the Union Square area is known to have a serious problem with theft, thus, not all mobile work places exhibit the same sense of security.

Community surveillance also plays a part in enabling people to be productive in mobile work places, although in a very different way. For example, Jackson, a freelance translator in Brooklyn who works at Starbucks daily from about 4 pm to midnight, finds that being surrounded by people – despite the fact that he doesn't know them – makes him less likely to 'goof off' and read Grokster, a gossip blog about New York. Instead, he focuses on his work and feels that he is getting more accomplished.

Social networks allow access to private information, diverse skills and power. Trust, diversity and brokers (or weak ties) are necessary to build powerful networks based on shared activities such as playing on a sport team, volunteering for a community organization and serving on a non-profit board (Granovetter, 1973; Uzzi and Dunlap, 2005). Mobile work places can be seen as hubs of information, skills and power, where everyone is a potential broker or weak tie. While informal interactions enable trust-building, the mutual recognition and the shared experience of working together day after day allows these informal interactions to become valuable for the exchange of private information, learning from one another and sharing access to new opportunities.

For example, on a particularly busy day in one of the world's busiest Starbucks, Victor, a self-employed 30-year-old graphic illustrator, was queuing for his ideal seat. While waiting in line, he began talking to Richard, a freelance web-designer and musician. Victor and Richard became friends and began working together on an almost daily basis. In the morning, the first person to arrive 'at work' would stake out space and notify the other by phone. If one Starbucks is too crowded, another coffee shop nearby is checked until an appropriate work place for the day is identified. Victor and Richard also met Daniel, the SEO in the narrative above, at Starbucks and work together with him every day. Victor and Richard have collaborated on several web-design projects, a sign that they have built trusting relationships that enable them to access new employment opportunities. In addition, Richard mentions that working alongside Victor and Daniel allows him to relieve stress more easily rather than becoming frustrated and giving up on his projects. Boundary-crossing and organizing diversity are elements that allow new media firms to remain innovative (Girard and Stark, 2002). Similarly, mobile work places foster the formation of social networks that bridge various clusters of knowledge. Rather than an isolated example, I found a number of mobile professionals who reported having developed social networks at mobile work places.

Any time? Anywhere?

Media representations of mobile work – in editorial coverage as well as advertising – focus on freedom, convenience and 'anytime, anywhere' access to mobile and wireless technologies. In contrast to these images, which reinforce convenience, freedom and ubiquity, mobile work places are sites of inconvenience, constraint and specificity. The following section illustrates the way in which, rather than being sites of 'anytime, anywhere' connectivity, mobile work places are deliberately chosen for specific purposes. For example, Victor had three regular mobile work places: a pre-production place, a production place and a deadline place. While all three places were Starbucks coffee shops, they were each unique in their relationship to the physical environment and to his social network. During the pre-production phase of his projects, Victor required books and materials that surrounded him at a Starbucks located within a Barnes & Noble bookstore in order to research the history, settings and characters for the storyboards that he was illustrating. However, these Starbucks were often smaller and had very few electricity outlets.

During the production phase of his projects, Victor moved to another Starbucks nearby where he could plug in his laptop and light-box (needed for tracing and drawing). It was at that Starbucks where he spent most of his time. His drawings were often spread out on the table and he had a constant stream of friends and visitors who knew that he worked there regularly. Finally, when he was on deadline, Victor went to a Starbucks in Korea-town. He didn't know anyone there and could work uninterrupted until he finished his project.

Victor chose each of these Starbucks, while seemingly identical to the average person, based on their unique physical, technological and social characteristics. While, for the most part, research on telecommuting and remote work assumes the elimination of commuting time in order to increase productivity and promote more sustainable transportation use (Gillespie and Richardson, 2000), Victor commutes 40 minutes to get to 'work' from his apartment in East New York (a poor neighborhood of Brooklyn). In addition, another informant, Jason, a remote technology salesperson for a Silicon Valley-based company, simulates his commute by taking a 30-minute walk to buy the newspaper or coffee on the days when he needs to be at home for private business phone calls before going to Starbucks for about four hours in the afternoon.

These examples illustrate the degree to which specific times and places are important to different people for varied reasons. As a regular patron of one Starbucks, Victor sometimes receives free coffee since he knows the Starbucks barista. He makes copies at the shop around the corner. He gets discounts on lunch nearby and likes to go out to dinner in the restaurants in the area after working from eight to 12 hours at the Starbucks. When asked why he commutes 40 minutes to get to work, he replies, 'Everything is *here*.' In fact, he got his first full-time job in the industry when an executive found him working on his drawings at 3 am at a Ray's Pizza in the East Village.

Conclusion

This chapter illustrates the ways in which mobile work places, and their unique social and digital ecologies, have become increasingly important in the lives of mobile professionals. Mobile work places support emergent occupations, such as search engine optimization, as well as occupations that have been transformed by information technology, such as graphic illustration and translation. Mobile professionals choose public places such as cafes, parks and other public and semi-public places for a wide variety of reasons. For example, some explain the need to separate their work life from their home life. Others seek the constant stimulation that public places provide or require the 'surveillance' of others in order to be (or to feel) productive.

Drawing on Carey's ritual view of communications, actor-network theory and Suchman's concept of situated action, I analyzed the interactions between people, technologies and places negotiated by WiFi hotspots. I found that mobile work places blur, and often reverse or contradict, traditional dichotomies such as work and play, online and offline, public and private, presence and co-presence, individual and community, and local and global. For example, in my study, I found that people often went outside the Lower East Side cafe, into very public places, to make private phone calls. In addition, in contrast to media representations of mobile work that focus on freedom, convenience and 'anytime, anywhere' access to mobile and wireless technologies, my year-long ethnographic study illustrates the ways in which WiFi hotspots enable local, face-to-face networks and communities.

I argue that mobile work places are important sites of informal interaction, social support and community. These factors are significant in promoting collaboration and innovation. As we shift from hierarchical forms of organizing to networked forms of organizing with the greater use of new media and information technology, it is vital to understand the ways in which mobile professionals rely on local, face-to-face communities encountered at WiFi hotspots. Mobile professionals rely on the community to stimulate their own productivity, provide surveillance for their personal belongings and computers, decompress after a long day, exchange ideas and identify new projects for collaboration.

This research illustrates a shift towards community forms of organizing and peer production, enabled by the complex interaction between emerging professions and WiFi technology as they are constituted in mobile work places. This study provides insight in order to inform scholars, managers, technologists, policymakers, architects and urban planners about emerging work practices. Mobile work practices are likely to increase in the near future due to the declining price of mobile and wireless technology as well as to the recent interest in cities around the world in building municipal wireless networks.

Notes

1 See www.nycvisit.com. Accessed on 1 August 2006.
2 See SurveyMonkey.com for more information.
3 See www.starbucks.com for more details. Accessed on 20 June 2007.
4 Spanish dance music – a blend of dancehall and hip hop – that was developed in the mid-1990s in Puerto Rico. See www.wikipedia.org. Accessed on 1 August 2006.
5 See www.nycvisit.com. Accessed 1 August 2006.

6 See www.citysearch.com. Accessed 1 August 2006.

7 In spring 2007, to the dismay of many longtime regulars, the owners of the cafe changed the layout, stopped serving food and limited the hours during which laptops and the wireless network could be used in order to cut costs and capitalize on their bar business.

References

Axinn, W. G. and Pearce, L. D. (2006). *Mixed Method Data Collection Strategies*. Cambridge: Cambridge University Press.

Benkler, Y. (2006). *The Wealth of Networks: How Social Production Transforms Markets And Freedom*. New Haven, CT: Yale University Press.

Bijker, W. E., Hughes, T. P. and Pinch, T. (1987). *The Social Construction of Technological Systems*. Cambridge, MA: MIT Press.

Bloemink, B., Hodge, B., Lupton, E. and McQuaid, M. (2006). *Design Life Now: National Design Triennial 2006*. New York: Smithsonian, Copper-Hewitt, National Design Museum.

Brown, B. and O'Hara, K. (2003). 'Place as a Practical Concern of Mobile Workers,' *Environment and Planning A*, 35: 1565–1587.

Carey, J. W. (1988). *Communication as Culture: Essays on Media and Society*. New York: Unwin Hyman, Inc.

Castells, M. (1996). *The Rise of the Network Society*. Malden, MA: Blackwell Publishers.

Creswell, D. J. W. (2003). *Research Design: Qualitative, Quantitative, and Mixed Methods Approaches*. London: Sage.

DeSanctis, G. and Monge, P. (1999). 'Introduction to the Special Issue: Communication Processes for Virtual Organizations,' *Organization Science*, 10/6: 693–703.

Gillespie, A. and Richardson, R. (2000). 'Teleworking and the City: Myths of Workplace Transcendence and Travel Reduction,' *Cities in the Telecommunications Age: The Fracturing of Geographies*, 228–245.

Girard, M. and Stark, D. (2002). 'Distributing Intelligence and Organizing Diversity in New Media Projects,' *Environment and Planning A*, 34/11: 1927–1949.

Granovetter, M. S. (1973). 'The Strength of Weak Ties,' *The American Journal of Sociology*, 78/6: 1360–1380.

Gupta, N. (2004). *Grande WiFi: Understanding What WiFi Users Are Doing in Coffee-Shops*. Cambridge, MA: MIT Press.

Horowitz, S., Buchanan, S., Alexandris, M., Anteby, M., Rothman, N., Syman, S. and Vural, L. (2005). *The Rise of the Freelance Class: A New Constituency of Workers Building a Social Safety Net*. New York: Freelancers Union.

Howard, P. (2002). 'Network Ethnography and the Hypermedia Organization: New Media, New Organizations, New Methods,' *New Media & Society*, 4/4: 550.

Innis, H. (1951). *The Bias of Communication*. Toronto: Canada University of Toronto Press.

Ito, M., Okabe, D. and Matsuda, M. (2005). *Personal, Portable, Pedestrian: Mobile Phones in Japanese Life*. Cambridge, MA: MIT Press.

Katz, J. E. and Aakhus, M. (eds) (2002). *Perpetual Contact: Mobile Communication, Private Talk, and Public Performance*. New York: Cambridge University Press.

Latour, B. (2005). *Reassembling the Social: An Introduction to Actor Network Theory*. Oxford: Oxford University Press.

Lee, S. (1999). 'Private Uses in Public Spaces,' *New Media & Society*, 1/3: 331–350.

Liff, S. and Laegran, A. S. (2003). 'Cybercafes: Debating the Meaning and Significance of Internet Access in a Cafe Environment,' *New Media & Society*, 5/3: 307–312.

Liff, S. and Steward, F. (2003). 'Shaping E-access in the Cybercafe: Networks, Boundaries and Heterotopian Innovation,' *New Media & Society*, 5/3: 313–334.

Mazmanian, M., Orlikowski, W. J. and Yates, J. (2006). 'CrackBerrys: Exploring the Social Implications of Ubiquitous Wireless Email Devices', European Group for Organizational Studies, unpublished paper.

Middleton, C. (2003). Broadband Internet Usage outside the Home: Insights from a Study of Toronto Internet Cafes. Ryerson University.

Moultrie, J., Nilsson, M., Dissel, M., Haner, U.-E., Janssen, S. and Van Der Lugt, R. (2007). 'Innovation Spaces: Towards a Framework for Understanding the Role of the Physical Environment in Innovation,' *Creativity and Innovation Management*, 16/1: 53–65.

Noam, E. (2005). 'The Commons as an Enabler of Commerce,' paper presented at the conference on The Economics of the Commons: Organizing Private Transactions in Communications, Columbia Institute for Tele-Information, New York.

Norman, D. A. (1990). *The Design of Everyday Things*. New York: Doubleday.

O'Mahony, S. C. (2002). *The Emergence of a New Commercial Actor: Community Managed Software Projects.* Stanford, CA: Harvard University Press.

Oldenburg, R. (1989). *The Great Good Place: Cafés, Coffee Shops, Community Centers, Beauty Parlors, General Stores, Bars, Hangouts and How They Get You through the Day*. New York: Paragon House.

Pedersen, P. E. and Ling, R. S. (2005). *Mobile Communications: Re-negotiation of the Social Sphere*. New York: Springer.

Poole, M. S. (1999). 'Organizational Challenges for the New Forms,' in G. DeSanctis and J. Fulk (eds), *Shaping Organizational Form: Communication, Connection, and Community*. London: Sage.

Powell, A. (2003). 'Space, Place, Reality and Virtuality in Urban Internet Cafes.' Ryerson and York Universities, unpublished paper.

Rheingold, H. (2003). *Smart Mobs*. Cambridge, MA: Perseus Books Group.

Sarbaugh-Thompson, M. and Feldman, M. S. (1998). 'Electronic Mail and Organizational Communication: Does Saying "Hi" Really Matter?,' *Organization Science*, 9/6: 685–698.

Schlosser, J. (2006). 'Cubicles: The Great Mistake,' *Fortune*, 22 March.

Stewart, J. (1999). *Cafematics: The Cybercafe and the Community*. Edinburgh: University of Edinburgh Press.

Suchman, L. (1987). *Plans and Situated Actions*. New York: Cambridge University Press.

Tashakkori, A. and Teddlie, C. (1998). *Mixed Methodology: Combining Qualitative and Quantitative Approaches*. Thousand Oaks, CA: Sage.

Thriving Office (2007). www.thrivingoffice.com.

Uzzi, B. and Dunlap, S. (2005). 'How to Build Your Network,' *Harvard Business Review*, 83/12: 53–60.

Wakeford, N. (2003). 'The Embedding of Local Culture in Global Communication: Independent Internet Cafes in London,' *New Media & Society*, 5/3: 379–399.

Watson-Manheim, M. B., Chudoba, K. M. and Crowston, K. (2002). 'Discontinuities and Continuities: A New Way to Understand Virtual Work,' *Information, Technology and People*, 15/3: 191–209.

Weiser, M. (1991). 'The Computer for the Twenty-First Century,' *Scientific American*, 265/3: 94–104.

4 Voluntary ghettos and mobile bureaucracy

Civic activity and acts of citizenship under threat

Tommy Jensen

Introduction: aim of chapter and author's ambitions

The aim of this chapter is to critically examine, first, the effects of the ongoing transformation of public spaces to accommodate mobile workers' needs, and, second, mobile workers' moral willingness and capability towards the Other. It is suggested that transformations of cities and the rising presence of mobile workers at public spaces are a potential threat to civic activity and citizenship acts. Let me also start this chapter revealing that my intention is to be provocative, somewhat speculative and intellectually curious.

Framing public spaces, civic activity and acts of citizenship

Public spaces are crucial for civic activity and acts of citizenship. Civic activity and acts of citizenship, following the insights of Hannah Arendt (1958/1998), can be depicted as the activity in which individuals have the urge, allowance and capability to express their opinions and to be recognized and confirmed. Public space is thus a potential place in which citizens jointly discuss and practice politics. Civic activity and acts of citizenship that take place in public space, or more relevant to this discussion the agora (a forum for citizens in ancient Greece), could also be depicted as an activity in which Oikos (the household) meets Ecclesia (an assembly open to all citizens of Athens). Cities have for a long period now been important economic, political and cultural nodes for the civilization process of mankind and, in late modernity, the economic, political and cultural importance of the city is enormous and still growing (Hardt and Negri, 2000). Importantly, cities are, and traditionally have been, important places in which definitions and the practice of democracy and citizenship are negotiated (Lidskog, 2006).

However, civic activity and acts of citizenship also concern how to collectively settle appropriate ways to discourage and eliminate injustice and to secure justice, fairness and solidarity. Consequently, the agora is an important place to discuss and to act on moral dilemmas. Elaborating on the agora and its importance to civic activity and acts of citizenship, we also need to consider, as Immanuel Kant notified us long ago (1793/1960), that humans by nature may be granted moral concern, but that this is something that needs to be taught (by one generation to another) and practiced (when being confronted by moral dilemmas). The agora is thus also important as a training ground for our willingness and capability to act morally responsible. In conclusion, the agora is a fundamental place in which 'human experience is formed and gleaned, life-sharing managed, its meaning conceived, absorbed and negotiated' (Bauman, 2003: 102).

However, civic activity and acts of citizenship are 'under siege' (Bauman, 2002). Metaphorically speaking, the core of the occupying force is capitalism, liberalism and economic globalization and the occupation weakens nation states and triggers departures from strategies intended to create, protect and improve social security to strategies concerning individual safety (Bauman, 1999). Interesting to this chapter is that these changes have severe negative consequences to the rich western urban areas and the denizens of its cities (but not as severe as to all those poor and vulnerable people living on our planet).

The negative consequences, due to the shift from social security to individual safety, which is particularly relevant to the context of modern city life can be depicted as follows: To live a modern western life is to live in an individualized society, where fear is endemic and daily lives are plagued by (1) threats to our body (growing old, substances in food, air and water, criminals and violence) and material possessions (robbery and burglary made by strangers and evil persons); (2) threatening changes to social security (related to income, employment, sickness and aging); and (3) threats to our identity and position in the social hierarchy (class, gender, sexuality, ethnicity, religion), and more generally the threat of social degradation and exclusion (Bauman, 2006).

In the individualized society, individuals feel that they are left on their own, and, to be able to secure a good life, they need to develop self-centred strategies (Beck, 1996). It is the most rational thing to do (albeit one-dimensional and bereft of moral content) in a world plagued by collective global problems and without any collective political institutions capable of solving them (Bauman, 1999). Cornelius Castoriadis has portrayed this as a condition in which private spheres – Oikos – invade and colonize public spheres – Ecclesia (Bauman, 2006). What before has been private matters, and thus something dealt with in privacy, is displayed publicly (soap operas, 'up-close and personal' with ordinary people and their confessions about problems, desires, perversions) but public and collective dilemmas are decentralized to be handled by (autonomous) individuals themselves (Bauman, 2002).

City ethics, mobile work and technology

According to the German moral philosopher Hans Jonas (1984), the manmade island of the city is the citadel in which all traditional ethics dwell. Particularly so, this apply to anthropocentric forms of ethics, invented in and for the artificial space of the city to ensure enclosure of man from nature (Jonas, 1984: 3–4). The foundation of city ethics is, Hans Jonas argues, predominately short-range ethics; moral concern about you and me, here and now. Although Jonasian ethics give rise to a number of challenges and interesting questions about man's moral concern for nature and future human generations (see Bonnedahl *et al.*, 2007 and Jensen, 2007 for critical suggestions on how to bring in non-anthropocentric and long-range ethics into the economy), the focus here is on moral concern between denizens that share the same city space (but that do not need to have a permanent residence).

Turning back to the first section, it can be concluded that the balance between Oikos, on the one hand, and Ecclesia, on the other, has been disturbed. Expressed differently, and adapted to the theoretical context suitable to this section, the foundations and ontology of city ethics are threatened, and at stake is our 'intimate immediacy for the nearest [and the] day-by-day sphere of human interaction' (Jonas, 1984: 6). The already short-range neighboring ethics of you and me, here and now, are getting shorter and shorter due to changes to (1) the architecture of physical places, (2) the actual use of places, and (3) the strategies to control and regulate places (cf. Jacobs, 1961/1993). However, as the third issue is out of reach for this chapter, only the first two issues will be elaborated upon.

To state that human bonds in cities are fragile is nothing new of course. Without the ambition to present a comprehensive list, central themes are (1) individualism and consumerism (self-centered, narcissistic acts to relieve desires and develop identities; consuming life as Bauman (2001a), would put it), (2) privatization of public spaces (private property, commercial business, gated communities), (3) the automobilization and motorscaping of space (cars, trucks, buses) to borrow phrases from Featherstone *et al.* (2005, see also Jane Jacobs 1961/1993), who prophetically warned us what could happen to cities in which the numbers and use of cars increase too much), and (4) technology in general and technological artefacts in particular (wireless connections to internet, smart systems to guide traffic flows, or monitoring techniques such as surveillance cameras).

A fifth important theme is mobile work and the use of mobile technologies in work. Recent innovations in information technology have radically paved new ways for mobile work (internet and the possibility of being constantly connected changes the means of production in a profound way). However, rising numbers of so-called knowledge workers (Alvesson, 2004), changes in lifestyle patterns and changing ways of organizing for- and non-profit organizations (the knowledge-intensive and flexibly managed organization) are important factors that also need to be brought into the analysis of mobile work. It is important to note, however, that, without information technology, we can hardly talk about mobility. Consequently, that is to say that knowledge, work, lifestyle patterns and organizing do not end up in mobility – without technology, we can only speak of flexibility.

Even if technology is granted a prominent position in understanding mobile work, most research primarily focuses on the functionality and capacity of technology to assist mobile workers. Moreover, research focus on possible benefits or downsides to the worker/s caused by changed ways of organizing (companionship, learning, sharing of knowledge and experiences – see Faulconbridge and Beaverstock, in this volume), and how a flexible and mobile working life collides with other spheres of life (such as family life and recreational activities – see Part IV, 'Home–work dynamics).'

What seems to be neglected is how developments of mobile work and mobile technology give rise to other types of problems. In the next section, I will outline what I refer to as two collective ethical dilemmas that mobile work and technology may give rise to: (1) changes made to the city landscape to accommodate mobile workers' needs, which amplify the ongoing process of constructing voluntary ghettos; and (2) since mobile work brings with it mobile bureaucracy, this has the potential to negatively influence the moral willingness and capability of mobile workers, to act and interact in public spaces.

Voluntary ghettos and mobile bureaucracy are both considered important, although the latter is more novel – and speculative – in character than the former. The collective ethical dilemma of voluntary ghettos primarily amplifies negative effects on civic activity and citizenship acts already observed in connection to the three first central themes outlined above (individualism and consumerism; privatization of public space; the automobilization and motorscaping of space). The collective ethical dilemma of mobile bureaucracy, however, is more freestanding and connected in a loose manner to theme one (individualism and consumerism) and four (technology).

Voluntary ghettos

Starting off with changes made to accommodate mobile workers' needs, it has for some time now been acknowledged that gradually cities and their public spaces are being transformed into sectors, or exclusive corridors, in which central areas of the city, i.e. fashionable and important areas, are reconfigured to fit consumers' needs. In the near future, however, the reconfiguration of cities to accommodate mobile workers' needs has the potential to accelerate dramatically. Consider the ongoing development of connected work places in air terminals, train stations, bus stations, subways, parks, squares, pavements, etcetera. A recent example is advertisements from corporations, providing internet access to urban consumers and mobile workers, announcing that it will soon be possible to access the internet in the 'open air' in central areas of cities (Stockholm was the city explicitly targeted here). What may not be apparent is that the increasing technological opportunities to scatter work through time and space (being a more flexible worker is a necessity in our contemporary society, being a mobile worker is a yet-to-come necessity) represent a threat to civic activity and acts of citizenship. In the process of adapting public space to mobile work and mobile workers' needs, an emerging notion of community, so far pushed forward by individualism and consumerism, privatization of public space and the automobilization and motorscaping of space, is further amplified. In this notion of community, 'Community means sameness, while 'sameness' means the absence of the Other' (Bauman, 2001b: 115). However, this notion of community is not only shared by denizens of cities; it is also strategically realized in practice. Accordingly, shared notions give rise to strategies, which in turn feed back to the notions themselves.

Zygmunt Bauman, when assessing the actual changes made, suggests that what we witness are constructions of voluntary ghettos. A voluntary ghetto is not a real ghetto, as for Bauman (2001b: 116–117) a decisive difference is that: "The real ghettos are places from which their insiders cannot get out ... the prime purpose of voluntary ghettos, on the contrary, is to bar outsiders from going in – the insiders are free to go out at will". Those inside real ghettos are thus denied freedom, Bauman continues, whereas those who voluntarily seek a socially closed environment and spatial confinement do so to achieve freedom. Voluntary ghettos, just as real ghettos, are territorial and social and achieve this by blending physical proximity/distance with moral proximity/distance, but closure and confinement need to be accompanied by a third element, the 'homogeneity of those inside contrasted with the heterogeneity of those outside...' (Bauman, 2001b: 116).

Physical and moral distance, sameness and homogeneity, seem to be the lodestar when cities undergo change. What emerges is a confined and closed agora, a context in which shared identities are assumed to exist a priori and where consensus is sought at all costs (Bauman, 2003). Such an agora is no good training ground for our willingness and capability to act in a morally responsible manner. On the contrary it generates mixophobic paranoia – fears of the unknown, strangeness – and paves the way for oppression:

> Social homogeneity of space, emphasized and fortified by spatial segregation, lowers in its residents their tolerance to difference and so multiplies the occasions for mixophobic reactions, making city life look more 'risk-prone' and so more agonizing, rather than making it feel more secure and so easier-going and more enjoyable.
>
> (Bauman, 2003: 113)

Contemporary city life looks more risk-prone to its denizens, Bauman argues. This is what the new notion of community, and the realization of voluntary ghettos strategies, brings with it! The urban existence seem to be plagued with contexts and situations that appear risk-prone, fostering needs of shelter and security, suspicion and avoidance. A perverse development considering that it is strangers that, by injecting new ideas and practices, initiate change to communities of people (Jacobs, 1961/1993). Furthermore, the attraction, pulse and nerve in city life stem from the possibility of being a stranger among other strangers (Bauman, 1993; 1995).

Leaving physical proximity and the negative changes that seem to have occurred in cities, I now turn to moral proximity and the interaction among denizens in order to explore the next troublesome ethical collective dilemma arising from mobile work.

Mobile bureaucracy

Cities adapting to mobility by providing and enhancing mobile space and mobile technology (such as wireless networks and specifically designed mobile work places) cause a change to the architectural settings, the actual use of public places, as well as to the types of denizens that are present. Negative ethical collective dilemmas due to changes made to city landscapes to accommodate consumers and their individual and self-centred pastimes (Bauman, 2001a; 2001c) or to accommodate certain places to the extraterritorial global capitalistic and working elites (airports and airplanes), busy establishing and maintaining social networks to stay at the centre of attention and power (Bauman, 1999; 2002), are much researched and discussed.

Here the focus is instead on ordinary public places and ordinary mobile workers (who work in knowledge-intensive organizations, but who do not belong to the capitalist and working elite). By emphasizing ordinary mobile workers working at ordinary public places, I will introduce the idea that not only workers and work-related activities are mobile; bureaucratic authority and discipline is also set free from its fixed organizational time-space configuration.

That workers are present and work is carried out in public places is, of course, nothing new (the meaning of the agora is also marketplace). The idea that workers are more or less disciplined by bureaucracy regardless of whether they are physically inside or outside organizations is not new either. In other words, bureaucracy follows the work activity, or activity of working, rather than being limited by corporeal or legal boundaries alone. What I wish to emphasize, and to which certain novelty could be ascribed, is, first, that the increasing numbers of mobile workers, carrying with them the mobile bureaucracy, have the potential to impact on day-to-day interactions among denizens in public places; and, second, that the mobility of bureaucracy has been dramatically increased by the carrying capacity of mobile technology.

Carrying out work, whether it is mobile work or not, is considered a threat to moral proximity and community relations because work is so radically different from civic activity and citizenship acts. When work is carried out, irrespective of location, contractual relations in authority structures, expectations of returns and non-personalized human behavior, in any form, prevail (Bauman, 2002). Obviously we could argue that this is especially so in for-profit organizations, but bureaucracy is always present, regardless of the context; sometimes its presence is strong, sometimes weak, but it is there (Weber 1964, cf. Jensen and Nylén, 2006). As Bauman (1989) argued so strongly: Even the Holocaust is an extreme but yet normal product of modernity and its prime invention, bureaucracy. Consequently, economical returns matters, but the moral dilemmas that may arise from mobile working can under no circumstances be reduced to this one dimension only.

We could say that work and bureaucracy have an inward direction (towards my career, my professional identity and my obligations to the organization) and civic activity and acts of citizenship have an outward orientation (towards the common good, the community and relationships with other citizens). Consequently, we could speak of different ethical foundations and how the mobile worker has difficulties breaking out of the equally mobile bureaucracy, because:

> [the] language of morality acquires a new vocabulary. It is filled with concepts like loyalty, duty, discipline – all pointing to superiors as the supreme object of moral concern and, simultaneously, the top moral authority. They all, in fact, converge: loyalty means performance of one's duty as defined by the code of discipline. As they converge and reinforce each other, they grow in power as moral precepts, to the point where they can disable and push aside all other moral considerations – above all, ethical issues foreign to the self-reproductory preoccupations of the authority system.
>
> (Bauman, 1989: 160)

To summarize, the ever-present effects of mobile bureaucracy on mobile workers means that, when they seem to be physically in a public place, they are morally not of that place (cf. Bauman, 2003: 98). Moral willingness and capability, as a duty, are not targeted towards those physically present here and now – the Other, the strangers – rather the here and now that matters is to be loyal to the bureaucratic regime and its superiors by fulfilling the set tasks.

Last, I will briefly turn the attention to mobile technology. The use of mobile technology, so ubiquitous these days, has the potential to morally disconnect people (an observation in stark contrast to the famous slogan 'connecting people'). Think of a person sitting on a bus, a train, a bench on the pavement as a café customer or in a park, concentrating on the screen of a laptop, or listening to music on an MP3 player, or talking intensely on a cellular phone, (see Elaluf-Calderwood and Sørensen, in this volume, for a description of how the use of mobile phones by passengers in London taxis means that they typically chat far less to the drivers).

If we merge the use of mobile technology with the duty to fulfil a specific work-related task, then the bureaucratic regime is not only present in the mobile workers' professional role and identity; it is also mediated by the technological gadget 'in hand' (cf. Jensen, 2004; 2006); a work-related text or numerical calculation that needs to be finished, a work-related conversation (with a voice on the phone; a face broadcast by video-link on the screen, email, SMS); or something to be learned by carefully listening to the MP3 player. If mobile bureaucracy, metaphorically speaking, manifests itself as a mask to secure loyalty, duty and discipline towards the bureaucratic regime, and thereby give rise to moral distance, then mobile technology, coming alongside mobile bureaucracy, could be depicted as a moral filter, also screening off those immediately present. That is to say that, even if physical proximity is shared, the use of technology might further amplify the moral distance that is set through mobile bureaucracy. This is also to say that physical distance cannot be overcome by technology (Bauman, 2002; Jonas, 1984).

A mobile worker is indeed at risk of screening him- or herself off from public interaction, but s/he might also screen off Others (who might want to initiate a thoughtful conversation, involving engagement and participation). To conclude, the use of mobile technology as such might hinder civic activity and acts of citizenship to some extent, if combined with a work-related task, and thus mobile bureaucracy, the negative collective ethical dilemma could be significant.

Public spaces, civic activity and acts of citizenship revisited

Returning to the concept of agora, which is neither private nor public – but both (Bauman, 1999), the first conclusion that could be drawn is that civic activity and acts of citizenship presuppose that strangers share space with strangers. The rise of voluntary ghettos is here depicted as a threat to this presupposition because it has at its foundation physical and moral distance. The second conclusion that could be drawn is that civic activity and acts of citizenship require closeness to the Other, whose face, in Levinas's (1969) terms, is always radically unique. I have depicted mobile bureaucracy, accompanied with mobile technology, as a threat – a mask and a filter – that might prevent close bonding between citizens.

Cities and city life need public places characterized by physical and moral proximity to the different and heterogeneous, i.e. a diverse mixture of people with different purposes that share space (Jacobs, 1961/1993), to fend off mixophobic reactions and to provide an agora in which civic activity and acts of citizenship can be the norm. Such an agora is characterized by people who offer:

> [a] standing invitation to meaningful encounter, dialogue and interaction [and that have] the ability to interact with strangers without holding their strangeness against them and without pressing them to surrender it or to renounce some or all of the traits that have made them strangers in the first place.
>
> (Bauman, 2001c: 27)

Jane Jacobs, while literally walking the streets of American big cities for a number of years, also reached the conclusion that diversity among people and (the architecture of) space is indispensable for a vital and healthy community, because people in these social settings develop the willingness to keep and capability of 'keeping the public spaces of the city safe, in handling strangers so they are an asset rather than a menace' (1961/1993: 532). But how is this moral willingness and capability developed? Arguably, the teaching and practicing of moral willingness and capability in the agora is key. How this moral willingness and capability is developed is described by Jacobs thus (1961/1993: 108):

> People must take a modicum of public responsibility for each other even if they have no ties to each other. This is a lesson nobody learns by being told. It is learned from the experience of having other people without ties of

> kinship or close friendship or formal responsibility to you take a modicum of public responsibility for you.

In the foreword to the 1993 edition of her book, Jacobs (1963/1993: xvi) also spoke of the city as an ecosystem with great similarities to natural ecosystems – the city's balance is vulnerable and fragile and easily disrupted or destroyed, and:

> A city ecosystem is composed of physical-economic-ethical processes active at a given time within a city and its close dependencies. [City ecosystems] require much diversity to sustain themselves.... The more niches for diversity of life and livelihoods ... the greater its carrying capacity for life.

The city needs mobile workers, no doubt about it. But too many concentrated in the same place at the same time could risk dispiriting others, if the mobile workers remain entrenched in the regime of mobile bureaucracy and technological gadgets. In fact, this may interrupt a vital part of the city's vulnerable ecosystem – the public places and its potential as an agora. If the balance is interrupted, then physical and moral distance among denizens might be the result. To keep the balance in the ecosystem, there need to be all kinds of people present and all kinds of social, cultural, residential and commercial activity too.

We should be careful when expressing worries about possible collective ethical dilemmas arising out of the presence of mobile workers and mobile technology in public spaces, i.e. the rise of voluntary ghettos and the intensified presence of mobile bureaucracy, so that we do not overestimate the dangers. However, when confronted with Bauman's (1989) account of 'Modernity and the Holocaust,' Arendt's (1963/1994) account of 'Eichmann in Jerusalem' and the 'Banality of Evil,' or Browning's (1998) account of 'Ordinary Men: Reserve Police Battalion 101 and the Final Solution in Poland,' which expose how weak our moral willingness and capability can be when confronted by the disciplining processes of authority, hierarchy, specialization, order-giving and order-taking, group pressure and career opportunities, we should not underestimate the possible collective ethical dilemmas either. Although the scale of the potential problem that I try to assess here, and the suffering inflicted on others, is much less serious than the events of the Holocaust revealed to us, the root causes and the nature of collective ethical dilemmas in voluntary ghettos and mobile bureaucracy are in essence the same – the dehumanizing effects (or the moral distancing) and the promotion of sameness and homogeneity (or physical distancing).

If a city needs the agora – as depicted here as a vital and sound public place – then we must also ask ourselves if we, because of the ongoing transformation of public spaces to accommodate, consuming commercial and mobile working activity, will witness a widespread geographical deportation of potential agoras to 'invisible' and marginalized parts of the city. I think this is a cause for concern and that the connection between mobility and power is troublesome.

This reconnects to an issue earlier stated as beyond the scope of this chapter: who manages to control and regulate city space? There is urgent need for scholars of different disciplines to investigate this matter.

References

Alvesson, M. (2004). *Knowledge Work and Knowledge-Intensive Firms*. Oxford, State (US): Oxford University Press.

Arendt, H. (1958/1998). *The Human Condition*. London: University of Chicago Press.

Arendt, H. (1963/1994). *Eichmann in Jerusalem: A Report on the Banality of Evil*. New York: Penguin.

Bauman, Z. (1989). *Modernity and the Holocaust*. London: Polity Press.

Bauman, Z. (1993). *Postmodern Ethics*. Malden, MA: Blackwell Publishing.

Bauman, Z. (1995). *Life in Fragments: Essays in Postmodern Morality*. Cambridge, MA: Basil Blackwell.

Bauman, Z. (1999). *In Search of Politics*. Cambridge: Polity Press.

Bauman, Z. (2001a). *The Individualised Society*. Cambridge: Polity Press.

Bauman, Z. (2001b). *Community*, Cambridge: Polity Press.

Bauman, Z. (2001c). 'Uses and Disuses of Urban Space,' in B. Czarniawska and R. Solli (eds) *Metropolitan Space and Discourse*, Malmö: Liber.

Bauman, Z. (2002). *Society under Siege*. Cambridge: Polity Press.

Bauman, Z. (2003). *Liquid Love: On the Frailty of Human Bonds*. Cambridge: Polity Press.

Bauman, Z. (2006). *Liquid Fear*. Cambridge: Polity Press.

Beck, U. (1996). *The Reinvention of Politics: Rethinking Modernity in the Global Social Order*. Cambridge: Polity Press.

Bonnedahl, K.-J., Jensen, T. and Sandström, J. (2007). *Ekonomi och Moral: Vägar mot Ökat Ansvarstagande* (Economy and Morality: Routes to Increased Responsibility). Malmö: Liber.

Browning, C.R. (1998). *Ordinary Men: Reserve Police Battalion 101 and the Final Solution in Poland*. New York: Harper.

Featherstone, M., Thrift, N. and Urry, J. (2005). *Automobilities*. London: Sage.

Hardt, M. and Negri, A. (2000). *Empire*. Cambridge, MA: Harvard University Press.

Jacobs, J. (1961/1993). *The Death and Life of Great American Cities*. New York: The Modern Library.

Jensen, T. (2004). Översättningar av konkurrens i ekonomiska laboratorier: Om ekonomiska teoriers förenkling, komplexitet och fördunkling i hälso- och sjukvården (Translations of Competition in Economic Laboratories: How Economic Theories are Simplified, Complex and Blurred within Health Care). Umeå: Umeå University.

Jensen, T. (2006). 'Fördunklad Organisering i en Heterogent Materiell Värld' (Blurred Organizing in a Material Heterogeneous World), in D. Ericsson (ed.) *Den Oavsedda Organisationen* (The Unexpected Organisation), Lund: Academia Adacta.

Jensen, T. (2007). 'Moral Responsibility and Sustainable Development: Jonasian Ethics for the Technological Age,' *International Journal of Innovation and Sustainable Development* 2, 1: 116–129.

Jensen, T. and Nylén, U. (2006) 'Striving for Spontaneity: Bureaucracy Strikes Back,' *International Research Journal Problems and Perspectives in Management*, 4, 2: 144–158.

Jonas, H. (1984). *The Imperative of Responsibility: In Search of an Ethics for the Technological Age*. Chicago, IL: University of Chicago Press.

Kant, I. (1793/1960). *Religion within the Limits of Reason Alone*, trans. T.M. Greene and H.H. Hudson. New York: Harper Torchbooks.

Levinas, E. (1969). *Totality and Infinity: An Essay on Exteriority*. Pittsburgh, PA: Duquesne University Press.

Lidskog, R. (2006). *Staden, Våldet och Tryggheten: Om Social Ordning i ett Mångkulturellt Samhälle* (City, Violence and Safety: Social Order in a Multicultural Society). Göteborg: Daidalos.

Weber, M. (1964). *The Theory of Social and Economic Organization*. New York: The Free Press.

Part II

Work-related travel

5 Travelling to work

A century of change

Colin G. Pooley

Introduction

For most adults the journey to work is a necessary part of everyday life undertaken without a great deal of thought. However, commuting can consume a substantial amount of time, and imposes obvious costs on the individual, their families, employers, society and environment. It can be estimated that in 2005 the average working adult in Britain spent some nine days per year travelling to and from work, a figure that has probably changed little since 1900.[1] Not only does this commitment reduce the amount of time available for other activities, but commuting itself also creates congestion, pollution and personal stress (Whitelegg, 1992, 1997; Docherty and Shaw, 2003; Adams, 2005). This chapter examines some of the ways in which travelling to work has altered in Britain over the past century, exploring both continuity and change. Overall, it argues that there has been substantial continuity in the ways in which people make decisions about travel to work, even though some aspects of the process have changed radically, and that an historical perspective can usefully inform contemporary transport policy.

Although the journey to work is superficially a simple concept, in practice it is much harder to define. It can be suggested that commuting forms part of a mobility continuum that extends from very short-distance everyday mobility at one end to long-distance residential migration at the other, with the opportunity for considerable overlap and blurring between categories (Pooley *et al.*, 2005). For instance, while some journeys to work are direct and involve no other activities, in many cases travel to or from work may include diversions to include tasks such as shopping, dropping or collecting children at school or nursery, calling on friends or to undertake leisure activities. Thus, in this sense, travel to or from work overlaps with, and becomes part of, everyday mobility. Some jobs themselves demand mobility, and thus travelling to work involves temporary relocation (for instance, to a hotel or rented flat). In this sense commuting and residential migration also become blurred. While for many the journey to work is fixed and routine, for others (for instance, building contractors) the place of work may vary on an almost daily basis. Thus the nature and length of the journey to work can vary considerably depending on where work is being undertaken. For others, travel to work may vary in a more regular way: perhaps they can work at home on certain days but need to be in an office on others. Moreover, many people have more than one job and thus any one individual may have several different journeys during a single week. Such changes have also allowed people to trade residential migration against commuting (for instance, opting for a longer journey to work instead of moving house following a job change) in a way that would not have been possible in the past (Pooley, 2003). It is argued that this conceptual complexity is important, because much traditional transport modelling assumes a repetitive, point-to-point commuting pattern, whereas the reality for many people may be much less simple.

Factors structuring the journey to work: what has changed?

Before examining available data on changes in travel to work over time, it is first necessary to assess some of the factors that structure commuting behaviour and how these have evolved. The most obvious changes relate to the availability and efficiency of transport technologies. Although almost all the forms of transport used today for commuting (trains, cars, trams, buses, bicycles) existed in some form a century ago, access for many people was severely restricted and some transport forms, at least, were substantially slower. Thus in 1903 there were just 17,000 licensed motor vehicles on the roads of Britain and the national speed limit was 20 mph. Licensed vehicles had risen to 6.3 million by 1956 and in 2005 to 32.3 million (with some 40.8 million licensed drivers (DVLA, 2003, 2005). Between 1930 and 1965 there was no national upper speed limit outside built-up areas for vehicles carrying less than seven people, and after that date a national upper limit of 70 mph, reduced to 60 mph on single carriageways in 1977 (DfT, 2004). Likewise, in relation to access, in 1900 the bicycle was still mainly used for leisure activities by an elite, whereas by the 1940s in Britain it was one of the most important forms of urban transport for men (Lloyd Jones and Lewis, 2000). In essence, the changes that occurred during the twentieth century were much less about new technologies, and much more about providing improved access to existing technologies (albeit often improved) which allowed people both to travel faster and gain more independence. The more detailed impacts of some of these changes on the journey to work are examined below.

However, many other factors also changed during the past century and, arguably, these had just as much impact on the journey to work as did new, improved or more widely available transport technologies. One key area of change has been in the nature of work and in workforce participation rates.[2] Whereas in 1911 83.8 per cent of the population aged ten years and over were in employment, in 2001 only 69.1 per cent of those aged 16–74 were employed. Thus one key change is that proportionately fewer people travel to work (though the actual number in employment has increased by over seven million people – an increase of 45 per cent on the 1911 figure). The most dramatic change has been in gender differences: whereas in 1911 83.8 per cent of men over ten were in employment, but only 32.5 per cent of women, in 2001 the comparable figures were 69.1 per cent and 56.7 per cent. Thus female workforce participation has more than doubled while, proportionately, male participation has reduced substantially (and in 2001 there were just 1.3 million more men in the workforce than had been the case in 1911). These figures obviously reflect well-established changes in economy and society with longer education, earlier retirement, better welfare provision supporting non-employment due to partial incapacity, and changes in both the aspirations of and opportunities for women. However, they do underline the fact that both the ubiquity and characteristics of those travelling regularly to work have changed substantially over the past century. In tandem with these trends, the British labour market has seen a reduction in factory-based work, and much greater employment in service and office sectors. For example, whereas in Manchester in 1841 some 55.8 per cent of employment was in the manufacturing sector and 26.1 per cent in the service sector, by 2001 the figures were 11.3 per cent and 73.5 per cent respectively (Vision of Britain, 2007). Together with the impact of successive employment legislation, this has allowed many people to work more flexibly, varying their hours and even days of work, and this has to some extent spread travel to work over a longer period of time, and has allowed employees to combine travel to work more effectively with other activities.

Other changes both in the structure of cities and the nature of society have also had an impact on the journey to work. The relative locations of both workplaces and homes have changed substantially over the past century. In 1900, most cities still retained a largely Victorian spatial structure with employment concentrated in the city centre and most people living in residential districts relatively close to their workplace. During the twentieth century much employment was decentralized, many people were able to move to more suburban locations (in both public and private housing estates) and travel to work became more complex, often involving difficult cross-town journeys rather than more simple movement from a suburb to the city centre. Improved transport, in part, allowed such trends but there were other, probably more important, economic forces at work. In combination, the changing structure of the city made travelling to work much more complex for many people and, in particular, meant that many fixed route options (trams, buses, trains) became less convenient (Dyos and Aldcroft, 1969; Pooley and Turnbull, 2000a). There have also been fundamental changes in society, with most people acquiring a wider range of commitments that have to be fitted into a finite amount of time. This process of time-space compression (Harvey, 1989) has, in turn, made many commuting journeys more complex as people link travel to or from work with other activities. Such factors often disproportionately affect women, who continue to take the greater responsibility for duties such as child care and shopping, despite workforce participation rates that almost equal those of men (Oakley, 1974; Coltrane, 1996). In conjunction with such trends, it can be argued that the twentieth century saw a fundamental change in people's mobility aspirations (Urry, 2000). Even if these aspirations have not always been fulfilled, new forms of media and communication have extended the possibility of mobility and have led to people expecting to be able to travel frequently and flexibly. In the past, it can be suggested, mobility aspirations were much more limited, but if new expectations are not realized, then this may lead to greater frustration with travelling (including the daily commute). Finally, it can also be suggested that, as people travel more, they view mobility in new ways. Whereas in the past travelling may have been seen as dead time, simply a method of getting from one place to another, in the twenty-first century it can be argued that travel is increasingly seen as an activity that is important for its own sake. Even the journey to work may be used productively (for instance, new mobile communications make it much easier to work while travelling) and thus the distinction between commuting time and working time may be increasingly blurred (Urry, 2000; Larsen *et al.*, 2006; Sheller and Urry, 2006).

Data

There are no data that provide a comprehensive overview of changes in the journey to work in Britain over the past century. In this chapter information is drawn from a number of public sources and from research completed by the author. National statistics on the journey to work mainly come from two sources: the census of population and the National Travel Survey (NTS). A question on travel to work was included for the first time in the 1921 census and this was repeated in 1951. From 1961 onwards there have been a small number of questions on commuting including information on transport mode. However, such data have only limited use for historical analysis. Census tabulations in 1921 and 1951 record only journey-to-work movement between local authority areas, thus ignoring the majority of flows that occur within an urban area, and they are presented only as broad aggregate statistics. They give a snapshot of longer-distance movement, but certainly do not reflect all commuting trips (Lawton, 1963, 1968; Warnes, 1972). The NTS supplies much more data for the recent past, and was designed specifically to collect information on daily travel patterns. The first NTS was carried out in 1965/6, with further ad hoc surveys through the 1970s and early 1980s, and with the collection of continuous survey data since 1988. The NTS provides detailed data, albeit drawn from a sample survey, on travel to work (together with most other aspects of mobility). However, though a valuable resource, it is not always possible to make direct comparisons over the full time period that surveys have been conducted, and the data are aggregated from information collected over randomly selected weeks. In addition to these statistics, a number of planning reports and related surveys provide historical data on commuting, especially in urban areas (see, for instance: Caradog Jones, 1934; Liepmann, 1944; Westergaard, 1957; Buchanan, 1963).

Other data used in this chapter were collected during 1996 as part of a project funded by the Leverhulme Trust. Information was collected, first, from a survey of 1,834 individuals scattered throughout Britain who began work after 1890 and who provided detailed information on their employment life histories, including 12,439 separate journeys to work; and, second, from 90 in-depth semi-structured interviews with respondents living in the case-study cities of London, Glasgow and Manchester/Salford. These interviews probed the structure and context of journey-to-work decisions. Respondents were identified in a number of ways. Our main source (providing 77.4 per cent of respondents) was a network of genealogists and family historians, which we had utilized in a previous project (Pooley and Turnbull, 1998), through which individuals provided information both about themselves and their immediate ancestors; but we also sought respondents by contacting large employers in selected towns, and by placing advertisements in the local press. These data, and the inevitable biases that they contain, have been fully discussed in other publications (Pooley and Turnbull, 1999, 2000b). This chapter makes selective use of all this material to examine both trends in the journey to work over time and the ways in which commuters made decisions about their daily journeys.

Contours of change

For most people travel time is much more important than distance. As access to faster forms of transport has increased, then most people have been able to commute further without extending the amount of time spent travelling (Schafer and Victor, 1997). This global trend is clearly illustrated by available UK data where, it is estimated, distance travelled on the journey to work has increased fourfold over the last century, but travel time has not even doubled (Table 5.1). Moreover, most of this increase in travel time occurred in the first half of the twentieth century, with mean travel times almost unchanged from 1940. This trend is further illustrated by consequent changes in the average speed of travel for the journey to work which, overall, more than doubled from the 1890s to the 1990s. The most recent data from the National Travel Survey suggest that there has been little change in either travel time or distance over the past decade with the average commuting trip remaining around 13–14 km and with a mean travel time of 25–27 minutes.[3] However, during the past ten years there has been a slight decline in both the number of commuting trips undertaken each year by the individuals sampled, and in the total distance travelled (DfT, 2006).

It is well established that there are consistent gender differences in most travel patterns, and this is especially true of commuting. As Table 5.1 shows, men have consistently travelled both further and faster than women, with the consequence that, for the first half of the century, on average women actually spent longer travelling to work than men even though female travel distances were shorter. There are also consistent variations between different locations, with commuters in London travelling both further and for longer than people elsewhere in Britain (Table 5.2). On average, the journey to work in London has been double that of elsewhere in the country for much of the last century though – due to a better public transport system – travel speeds in London have been mostly a little faster than elsewhere. The dominance and distinctiveness of

Table 5.1 Average distance (km), time (min) and speed (km per hour) travelled for journeys to work since 1890 by gender

Decade	*Males*			*Females*			*All*		
	Distance	*Time*	*Speed*	*Distance*	*Time*	*Speed*	*Distance*	*Time*	*Speed*
1890–9	4.0	17.0	14.1	1.8	21.3	5.1	3.6	17.7	12.2
1900–9	3.9	21.5	10.9	3.2	25.4	7.6	3.8	22.4	10.2
1910–19	6.2	27.0	13.8	5.1	26.8	11.4	5.9	27.0	13.1
1920–9	6.8	28.2	14.5	6.1	31.3	11.7	6.7	29.0	13.9
1930–9	7.0	30.5	13.8	6.8	31.9	12.8	7.0	30.9	13.6
1940–9	8.2	33.8	14.6	7.3	33.1	13.2	7.8	33.5	14.0
1950–9	10.1	33.6	18.0	7.4	34.4	12.9	9.0	33.9	15.9
1960–9	12.1	34.6	21.0	7.5	32.1	14.0	10.2	33.5	18.3
1970–9	13.1	34.5	22.8	7.6	28.5	16.0	10.3	31.5	19.6
1980–9	15.5	37.3	24.9	8.8	29.4	18.0	12.0	33.1	21.8
1990–8	19.4	39.1	29.8	10.5	30.7	20.5	14.6	34.5	25.4

Source: Details of 12,439 journeys to work taken from 1,834 individual life histories. Statistics relate to all modes of transport and are calculated for the decade in which a particular journey to work started.

commuting to London has been a consistent trend throughout the twentieth century. For instance, in 1921 daily in-movement to London was almost four times that of the city with the second largest in-flow of commuters (Manchester) and, although this differential reduced in mid-century (with London the only major city to experience a decline in daily in-movement between 1921 and 1951), even by the later date daily commuting to London remained more than double that of in-movement to Manchester (Lawton, 1963).

Most of the trends outlined above can be explained by variations in the mode of transport used to travel to work. Whereas a century ago approximately half of all journeys to work were undertaken on foot, with most of the rest by a variety of forms of public transport; by mid-century walking accounted for only about one-fifth of all commuting, with a further fifth by bicycle and less than 10 per cent by car or van; and by the late twentieth century over half of all commuting was undertaken in a private vehicle with around 30 per cent of journeys by public transport (Table 5.3). According to the 2001 census in England and Wales just 10 per cent of people normally travel to work on foot and 2.8 per cent by bicycle. Again, there are substantial variations by gender and location (Tables 5.4, 5.5). Women have been consistently more likely than men to walk or take public transport, and it is only in the last decade that male and female car use has become almost equal (though women are still more likely than men to be passengers rather than car drivers). Locational differences reflect variations in transport provision and, most recently, in the degree to which the private car has been regulated. Public transport provision has always been much better, and use much higher, in London than elsewhere, and this trend has been further enhanced by the introduction of congestion charging in the capital (GLA, 2004).

Table 5.2 Average distance (km), time (min) and speed (km per hour) travelled for journeys to work since 1890 by location of workplace

Workplace	*Time period*				
	1890–1919	*1920–39*	*1940–59*	*1960–79*	*1980–98*
London					
Distance	6.8	11.1	14.3	18.0	20.5
Time	29.0	43.3	50.7	52.2	51.5
Speed	14.1	15.4	16.9	20.7	23.9
Other cities >100,000 population					
Distance	4.3	5.6	6.5	8.3	10.2
Time	25.3	27.4	28.8	29.3	30.3
Speed	10.2	12.3	13.5	17.0	20.2
Towns <100,000 population					
Distance	3.7	4.4	6.4	7.9	10.9
Time	20.2	21.7	26.4	25.1	26.1
Speed	11.0	12.2	14.6	18.9	25.1
All locations					
Distance	5.0	6.8	8.5	10.2	12.8
Time	24.7	30.3	33.7	32.6	33.6
Speed	12.2	13.5	15.1	18.8	22.9

Source: Details of 12,439 journeys to work taken from 1,834 individual life histories. Statistics relate to all modes of transport and are calculated for the decade in which a particular journey to work started.

Bicycle use and walking are consistently most common in smaller communities, though cycling has also recently increased in London, but car use is also high in small towns and rural areas. Thus, according to the 2001 census, in the city of Liverpool 55.1 per cent of people travel to work as a car driver or passenger and 12.3 per cent either walk or cycle but, in the much smaller town of Lancaster, while 62.5 per cent travel to work in a car, some 18.2 per cent of commuters also walk or cycle.

According to the limited available evidence, the reasons why people choose a particular form of transport have changed relatively little over time (Table 5.6). In almost all time periods, and for both men and women, the main stated reason was lack of choice, followed by personal preference. However, the dominance of both lack of choice, and of cost constraints, has declined over time, with personal preference becoming increasingly more significant. This indicates that travel constraints have become gradually less severe as people have been able to choose the mode of travel to work that suits them best. There were relatively few variations by gender or location in the reasons given for choosing a particular mode of transport to travel to work: women have been consistently more concerned than men about speed – probably indicating the greater time constraints that women operate under and the fact that for the most part they use slower forms of transport than men – and speed of travel was also more

Table 5.3 Main mode of transport for journeys to work in Britain since 1890 (%)

Decade	*Transport mode*							
	Walking	*Bicycle*	*Tram/trolley bus*	*Bus*	*Train*		*Motor cycle*	*Car/van*
					(overground)	*(underground)*		
1890–9	59.4	2.0	16.8	5.0	9.8	5.0	0.0	0.0
1900–9	49.4	11.2	11.6	14.6	10.2	0.4	0.0	1.1
1910–19	40.6	13.3	16.0	9.9	15.4	1.9	0.6	1.9
1920–9	28.5	17.5	10.6	15.3	17.8	2.3	3.9	5.2
1930–9	22.5	19.1	9.7	13.8	18.4	4.1	2.3	9.1
1940–9	17.2	19.6	6.7	23.0	18.3	5.4	2.2	6.0
1950–9	13.4	16.0	2.5	23.3	18.9	4.4	3.0	16.3
1960–9	14.0	5.2	0.2	18.8	16.2	5.3	2.6	35.8
1970–9	13.4	4.5	0.1	15.8	13.2	5.5	1.9	44.5
1980–9	10.3	6.1	0.0	11.7	15.4	5.4	1.8	48.5
1990–8	7.9	6.1	0.2	10.1	17.2	4.5	0.6	52.8
Sample size	2,083	1,379	466	2,073	2,002	564	264	3,108

Source: Details of 12,439 journeys to work taken from 1,834 individual life histories. Statistics are calculated for the decade in which a particular journey to work started.

Table 5.4 Main mode of transport for journeys to work since 1890 by location of workplace (%)

Mode	*Time period*				
	1890–1919	*1920–39*	*1940–59*	*1960–79*	*1980–98*
London					
Walk	28.7	11.5	6.6	4.7	3.4
Bicycle	5.2	4.0	5.6	1.5	5.8
Tram/trolley bus	14.9	9.5	3.7	0.1	0.0
Bus	20.1	20.6	16.2	10.2	12.5
Train (overground)	24.2	38.9	44.7	45.7	41.8
Underground	4.8	9.9	17.8	23.4	21.8
Motor cycle	0.0	1.2	1.3	1.6	1.7
Car/van	0.7	4.2	3.5	12.5	12.7
Sample size	289	506	954	792	464
Other cities >100,000 population					
Walk	46.6	25.6	13.4	12.3	12.2
Bicycle	11.2	19.6	18.2	5.2	5.3
Tram/trolley bus	26.1	18.4	8.1	0.2	0.1
Bus	7.5	11.6	31.2	27.6	15.4
Train (overground)	7.5	13.1	10.6	7.6	12.7
Underground	0.0	1.1	0.2	0.7	0.4
Motor cycle	0.4	2.9	2.8	2.5	0.9
Car/van	0.7	6.2	14.2	42.9	52.1
Sample size	268	550	1,316	1,260	755
Towns <100,000 population					
Walk	61.9	34.4	22.2	19.1	10.3
Bicycle	17.5	29.0	24.9	6.2	6.9
Tram/trolley bus	3.8	3.4	1.6	0.1	0.0
Bus	4.5	10.2	20.5	13.0	6.9
Train (overground)	7.0	6.3	8.9	5.4	5.3
Underground	0.4	0.3	0.7	0.4	0.5
Motor cycle	0.7	4.2	3.5	2.3	1.6
Car/van	2.8	11.6	14.7	50.8	67.5
Sample size	286	649	1,478	1,651	889

Source: Details of 12,439 journeys to work taken from 1,834 individual life histories. Statistics are calculated for the decade in which a particular journey to work started.

important in London than elsewhere, no doubt reflecting the longer period of time spent commuting in the capital. If the daily commute is viewed within the context of all everyday mobility then one other trend also becomes apparent. Whereas in 1965 work-related travel accounted for almost 40 per cent of all trips, by 2001 this had declined to just 18.7 per cent (DoT, 1979; DfT, 2001). Thus it can be suggested that, in the past, everyday mobility was dominated by essential movement related to work and other necessary activities, but in the twenty-first century, although most people in employment do still travel to work daily (according to the 2001 census some 9.2 per cent of people in employment

Table 5.5 Main mode of transport for journeys to work since 1890 by gender (%)

Mode	*Time period*				
	1890–1919	*1920–39*	*1940–59*	*1960–79*	*1980–98*
Males					
Walk	44.3	23.2	12.3	8.0	7.2
Bicycle	13.2	21.6	21.4	5.5	7.1
Tram/trolley bus	13.8	9.0	3.3	0.0	0.0
Bus	9.8	10.8	17.7	11.3	5.8
Train (overground)	14.4	17.5	18.9	17.9	21.9
Underground	1.2	3.2	3.8	3.9	4.0
Motor cycle	0.5	3.5	3.9	3.3	1.8
Car/van	1.6	10.0	16.3	48.1	50.9
Sample size	646	1,272	2,291	2,026	1,000
Females					
Walk	50.3	29.3	19.7	20.6	11.6
Bicycle	5.0	9.5	11.7	4.0	5.1
Tram/trolley bus	17.8	13.2	6.2	0.2	0.1
Bus	14.2	22.2	31.7	24.8	16.1
Train (overground)	8.6	20.3	18.1	11.1	10.1
Underground	3.6	3.9	6.6	7.2	6.1
Motor cycle	0.0	0.9	0.6	1.1	1.0
Car/van	0.5	0.7	4.4	29.9	49.0
Sample size	197	433	1,457	1,677	1,108

Source: Details of 12,439 journeys to work taken from 1,834 individual life histories. Statistics are calculated for the decade in which a particular journey to work started.

work mainly from home), non-essential movement for leisure and social activities (including shopping) has become increasingly significant.

The experience of change: Britain *circa* 1930–*circa* 1970

It can be argued that the key period of change in commuting in Britain was from the 1930s to the 1970s. During this period, not only did the private car become increasingly available, but there were also significant changes in the structure of economy and society, and in the physical structure of urban areas, which affected the journeys that people undertook and the range of travel choices that they had. In this section changes in the journey to work are examined through the eyes of respondents who lived through this period in the three cities of Glasgow, Manchester and London, focusing on four main factors which, it can be suggested, affected individual decisions about everyday travel: cost, travel time, convenience and comfort. In this chapter only brief examples can be given but full details of the data can be found in Pooley *et al.*, 2005.

For many respondents cost was a major factor that encouraged them to walk to work. In the 1930s ands 1940s many families had very limited incomes and it

Table 5.6 Reasons for choosing a particular mode of transport for the journey to work since 1890 by gender (%)

Reason	*Time period*					
	1890–1919	*1920–39*	*1940–59*	*1960–79*	*1980–98*	*1890–1998*
Males						
Lack of choice	60.4	41.3	41.4	40.8	38.0	42.2
Personal preference	21.1	31.0	32.5	36.3	38.1	33.3
Cost	9.1	13.0	10.3	4.7	5.9	8.4
Speed	7.0	11.3	11.5	11.2	10.7	10.9
Transport provided	2.4	3.4	4.3	7.0	7.3	5.2
Sample size	497	1,065	2,074	1,849	964	6,449
Females						
Lack of choice	59.6	40.2	46.9	41.9	39.2	43.3
Personal preference	18.0	27.3	27.8	32.9	35.0	31.0
Cost	13.1	9.8	5.0	3.1	2.8	4.6
Speed	9.3	17.9	17.6	16.2	16.9	16.6
Transport provided	0.0	2.8	2.7	5.9	6.1	4.5
Sample size	161	358	1,293	1,534	1,019	4,365

Source: Details of 12,439 journeys to work taken from 1,834 individual life histories. Statistics are calculated for the decade in which a particular journey to work started.

made sense to leave home a little earlier and walk to work rather than spend money on a tram or bus fare. Cost constraints seem to have been especially important for female respondents – who were possibly saving money to spend on food or other necessities – while male respondents were much less likely to cite the need to save money as a reason for walking to work. Two respondents from Manchester were typical:

> I soon decided to get up a bit earlier and walk because … I was paying … half a crown for tram fares, and I thought I can't even save up to go home. So I started walking to work and walking back.
>
> (RJ16, Manchester, female, 1930s)

> It was financial. If I walked I was saving a penny/penny halfpenny you know, what I would have spent on the fare I could spend.
>
> (RJ17, Manchester, female, 1940s)

Male respondents who expressed concern about the cost of travelling to work in mid-century were much more likely to cycle than to walk. This could reflect the fact that, on average, men travelled further to work than women (and thus walking was less practical), but it must also relate to the fact that many more men than women had access to a bicycle. Indeed, in many areas by the 1930s and 1940s the bicycle was the single most important means of travel to work for men. One London respondent from the 1950s was typical:

> when I started I didn't think I earned very much and by the time I'd paid, well fare to work, I thought this is not on. So if I went on my bike it wouldn't cost me anything.
>
> (RJ87, London, male, 1950s)

For many respondents travel by public transport was also viewed as relatively cheap, certainly much more economical than running a car. Some commuters varied between walking and the bus or tram depending on weather conditions or other factors, and even those who had access to a car rarely used it for work before the 1960s due to a combination of cost and convenience. These views were expressed by a variety of respondents:

> Well tram cars was … the mode of transport.… That was the normal mode of transport and it was very cheap in these days. That was the only mode of transport … You'd no option.
>
> (RJ49, Glasgow, male, 1930s)

> In 1955 the area where I lived … was quite a poor area. I mean we didn't have then what we've got now and … no-one had a car either. Then, the best facility to get me from like the dock area to the main station, tube station, was a bus. There was no other choice, or walk, which I did do quite often.… There was no other choice.
>
> (RJ91, London, male, 1950s)

> If you had access to a car at that stage … you would have used that for leisure only. It would not have occurred to you to use it for work.
>
> (RJ04, Manchester, male, 1950s)

Concepts of travel time, convenience and comfort are all closely linked together, and interacted with cost to determine the form of transport that people used to travel to work. For many men in the 1930s and 1940s the bicycle offered the ultimate form of quick and convenient transport. Cycling to work provided independence and freedom to choose a route; it did not require interaction with other travellers and was often quicker than travelling by public transport. Sometimes, especially when on shift work, poor public transport provision meant that cycling was the only viable option. It can be suggested that the reasons given (below) by male respondents who chose to cycle to work in mid-century are very similar to the reasons why people travel by car today (Sheller and Urry, 2000; Beckmann, 2001; Urry, 2004). The combination of privacy, flexibility and speed has always been important and by the 1930s this could be achieved most easily by cycling.

> Well it [cycling] was really the only way. 'Cause there was such a tremendous detour using public transport … well the time factor, it was horrendous.
>
> (RJ14, Manchester, male, 1930s)

> I'm afraid I never liked travelling on trams ... they were never very comfortable and if you went on the top deck it was very uncomfortable because there was smoking on the upper deck. I would sooner ride a bike.
>
> (RJ32, Manchester, male, 1930s)

> Well I used to get up at about, I used to get up at about six o'clock I think and I would cycle, it was the only way, there was no other way of doing it ... there was no public, there was no buses at that time. I had to cycle to the station which was about four miles.
>
> (RJ82, London, male, 1950s)

Most respondents did not express strong views about the comfort and convenience of public transport: they simply accepted that it was the normal way in which people travelled. In most cities in the 1930s and 1940s trams and buses were the taken-for-granted method of commuting. Objections to public transport came mainly from men, and mostly related to the inconvenience of having to travel with other people to a predetermined timetable. In contrast some women expressed quite positive views about the sociability of travelling by public transport: a view that was still expressed by some respondents travelling in the 1990s.

> I didn't have any other form of transport and very few people, in those days anyway, had cars, particularly coming from the council estate where I did, it was a rarity. So, I mean, everyone travelled by bus whether it was to work or social ... or whatever. I mean it was just a natural thing, everyone automatically went by bus.
>
> (RJ64, Manchester, male, 1960s)

> The days I did have to go on the bus I used to get extremely annoyed and, that was if it was raining, and it, it put me out for the whole day.... I'd got, just an obsession, that I was not going to travel on a [bus].
>
> (RJ35, London, male, 1930s)

> And I think really I, I enjoyed the bus journeys because I could read. I always read on bus journeys.
>
> (RJ85, London, female, 1970s)

> I've got quite used to getting the bus now and I've, I've met a few people, who I've known through getting the bus with, and I quite like having a chat with them in the morning.
>
> (RJ65, Manchester, female, 1990s)

By the 1950s and 1960s an increasing number of (mainly male) respondents had access to a car and were using it to travel to work. Some of the main factors that stimulated the use of a private car for commuting (as opposed to leisure) were the provision of the company car and the changing location of homes and workplaces as travel to work increasingly involved complex cross-city journeys that could not easily be accomplished by public transport. However, some respondents (especially women) continued to find commuting by car stressful and unattractive, and preferred to take public transport.

> Yes, I got a car at that point because to travel to x was quite awkward. To do it by public transport would mean … a bus journey, an underground journey, and another bus journey … so it really wasn't terribly convenient, so I'd managed to accrue a little capital and I bought a car.
>
> (RJ39, male, Glasgow, 1950s)

> I went to work by car. Very often you use the car an awful lot because … you're repeatedly going out to firms … and you were paid so much for using your car … so you see you had a good reason for it really.
>
> (RJ08, London, male, 1950s)

> I did think of learning, I think I did try at some point, but I just really, really hated it and just didn't want to do it. Just, just didn't like driving.
>
> (RJ67, Manchester, female, 1960s)

While the main attraction of walking was financial, respondents also cited other reasons for choosing this means of travelling to work. Some simply enjoyed being in the fresh air, but for others – even in the 1930s – walking was a means of transport that was only used when there was nothing else available.

> Very pleasant, down through houses, residential. Could have got a bus but I preferred to walk cause I like walking.
>
> (RJ04, Manchester, Male, 1950s)

> There were times … when there were very bad fogs. Smogs, smoke and fog you know. And though the tram cars went on lines they did sometimes go off, and you just had to walk home from work you know, that was it.
>
> (RJ22, Glasgow, male, 1930s)

> Well, again, I think I've probably had to walk home in snow or thick fog or something like that, but it wasn't from choice.
>
> (RJ37, Glasgow, female, 1940s)

Conclusions: implications for policy

It might be assumed that the ways in which people travelled to work in the past have little relevance for present-day transport policy. However, it can be argued that understanding how commuting decisions have changed may help planners and policy makers to appreciate the constraints under which many people travel, and the likely impact of policy measures. Analysis of changes in the journey to work in Britain over the past century has demonstrated that travel time is much more important than travel distance; that commuters (especially men) have always tended to opt for the form of transport that offers most independence and flexibility; that all travel decisions result from an interaction of preferences and constraints (including, but not necessarily primarily, cost); and that there are consistent gender differences in the ways in which men and women travel to work. These trends might suggest that commuting patterns and preferences are quite firmly established and that policy decisions that attempt to prise commuters out of cars and onto public transport or bicycles are likely to have little effect. However, it can also be argued that in the past the travelling public have had to adapt to substantial changes in the transport infrastructure, and have proved able to change their everyday travel accordingly. Thus when trams were removed from the streets of most British towns, people accepted this without difficulty and rapidly switched to other means of transport, as stated by one respondent from Glasgow:

> Well there was widespread news that they were, they were stopping the tram, the trams. We were all in advance by it, well advanced warning about buses taking their place.... Well it was, it was well known that they were going to change trams to buses so it was well publicized.
>
> (RJ22, Glasgow, male, 1960s)

If people have changed their travel behaviour in the past, then they can do so in the future, suggesting that (despite some initial opposition) schemes to introduce road pricing or congestion charging can lead to real behavioural change. This has certainly been the experience in London where, after the introduction of congestion charging in 2003, traffic reduced by some 15 per cent during congestion charging hours (GLA, 2004). However, what is also clear from accounts of journeys in the past is that people have always been concerned about the quality of their travelling environment, and that trips that provide enjoyment and/or enable people to engage in other activities have always been preferred. Thus, it can be suggested that, while for many people the opportunities for mobility are greater than ever before, and that travel experiences are increasingly diverse, there has been much less change in the journey to work than in some other travel, and that forms of transport that combine convenience, low cost and a congenial environment have always been preferred.

Notes

1 The 2005 figure is based on an average one-way commuting trip of 27 minutes (DFT, 2006) over a five-day working week and 48-week working year. The figure for 1900 is based on an average one-way commute of 22 minutes over a six-day working week and 50-week working year (see Pooley and Turnbull, 1999, 2000a, 2000b for details).

2 It is not possible to compute directly comparable figures for labour force participation. Data are from Census of England and Wales, 1911 and 2001.

3 The mean travel distance in the NTS survey and the sample data used in Table 5.1 are very similar, but NTS data show somewhat lower travel times. This could be due to the

additional NTS category of business trips (which could include some commuting and which are on average longer) or to a slight over-representation of London in the time-series data.

References

Adams, J. (2005). 'The Limits to Integration: Hypermobility: A Challenge to Governance', in C. Lyall and J. Tait (eds), *New Modes of Governance: Developing an Integrated Policy Approach to Science, Technology, Risk and the Environment.* Aldershot: Ashgate, pp. 123–38.

Beckmann, J. (2001). 'Automobility – A Social Problem and Theoretical Concept', *Environment and Planning D: Society and Space*, 19: 593–607.

Buchanan, C. (1963). *Traffic in Town.* London: HMSO.

Caradog Jones, D. (1934). *The Social Survey of Merseyside.* Liverpool: University of Liverpool Press.

Coltrane, S. (1996). *Family Man: Fatherhood, Housework and Gender Equity.* New York: Oxford University Press.

DfT (Department for Transport) (2001). *National Travel Survey: 1999/2001 Update.* London: National Statistics.

DfT (2004). *Speed: Know your Limits.* London: DfT.

DfT (2006). *Transport Statistics Bulletin: National Travel Survey 2005.* London: DfT.

Docherty, I. and Shaw, J. (2003). *A New Deal for Transport?* Oxford: Blackwell.

DoT (Department of Transport) (1979). *National Travel Survey.* London: DoT.

DVLA (Driver and Vehicle Licensing Authority) (2003). Press Release 3, February 2003. Swansea: DVLA.

DVLA (2005). *Annual Report 2004–5.* Swansea: DVLA.

Dyos, H. J. and Aldcroft, D. (1969). *British Transport: An Economic Survey from the Seventeenth Century to the Twentieth.* Leicester: Leicester University Press.

GLA (Greater London Authority) (2004). *The Mayor's Transport Strategy Revision.,* London: GLA.

Harvey, D. (1989). *The Condition of Postmodernity.* Blackwell: Oxford.

Larsen, J., Urry, J. and Axhausen, K. (2006). *Social Networks and Future Mobilities.* Aldershot: Ashgate.

Lawton, R. (1963). 'The Journey to Work in England and Wales: Forty Years of Change', *Tijdschrift voor Economische en Sociale Geographie*, 44: 61–9.

Lawton, R. (1968). 'The Journey to Work in Britain: Some Trends and Problems', *Regional Studies*, 2: 27–40.

Liepmann, K. (1944). *The Journey to Work: Its Significance for Industrial and Community Life*. London: K. Paul, Trench, Trubner.

Lloyd Jones, R. and Lewis, M. (2000). *Raleigh and the British Bicycle Industry: An Economic and Business History 1870–1896*. Aldershot: Ashgate.

Oakley, A. (1974). *The Sociology of Housework*. London: Robertson.

Pooley, C. (2003). 'Mobility in the Twentieth Century: Substituting Commuting for Migration', in D. Gilbert, D. Matless and B. Short (eds), *Geographies of British Modernity*, Oxford: Blackwell, pp. 80–96.

Pooley, C. and Turnbull, J. (1998). *Migration and Mobility in Britain since the Eighteenth Century.* London: UCL Press.

Pooley, C. and Turnbull, J. (1999). 'The Journey to Work: A Century of Change', *Area*, 31: 282–92.

Pooley, C. and Turnbull, J. (2000a). 'Commuting, Transport and Urban Form: Manchester and Glasgow in the Mid-Twentieth Century', *Urban History*, 27: 360–83.

Pooley, C. and Turnbull, J. (2000b). 'Modal Choice and Modal Change: The Journey to Work in Britain since 1890', *Journal of Transport Geography*, 8: 11–24.

Pooley, C., Turnbull, J. and Adams, M. (2005). *A Mobile Century?: Changes in Everyday Mobility in Britain in the Twentieth Century*. Aldershot: Ashgate.

Richardson, T. (ed.) (2004). 'Interface. Planning and the Big C: Challenging Auto Dependence through Conviction Politics in London', *Planning Theory and Practice*, 5: 487–514.

Schafer, A. and Victor, D. (1997). 'The Past and Future of Global Mobility', *Scientific American*, October: 36–9.

Sheller, M. and Urry, J. (2000). 'The City and the Car', *International Journal of Urban and Regional Research*, 24: 737–57.

Sheller, M. and Urry, J. (2006). 'The New Mobilities Paradigm', *Environment and Planning A*, 38: 207–26.

Urry, J. (2000). *Sociology beyond Societies: Mobilities for the Twenty-first Century*, London: Routledge.

Urry, J. (2004). 'The "System" of Automobility', *Theory, Culture and Society*, 21: 25–39.

Vision of Britain through Time, www.visionofbritain.org.uk/index.jsp (accessed June 2007).

Warnes, A. (1972). 'Estimates of Journey to Work Distances from Census Statistics', *Regional Studies*, 6: 315–26.

Westergaard, J. (1957). 'Journeys to Work in the London Region', *Town Planning Review*, 28: 37–62.

Whitelegg, J. (1992). *Traffic Congestion: Is There a Way Out?* Hawes: Leading Edge.

Whitelegg, J. (1997). *Critical Mass: Transport, Environment and Society in the Twenty-first Century*, London: Pluto Press.

6 The business of train travel

A matter of time use

Glenn Lyons, David Holley and Juliet Jain

Introduction

It is evident to the casual observer travelling on the British rail network that fellow travellers are occupying their time with a range of activities – including work. The significance of such time use has tended to be overlooked in transport studies and in the economic assessment of time spent travelling and investment in measures to save such time. In transport studies, travel has traditionally been seen as a derived demand (Tipping, 1968) – derived from the need or desire to participate in activities that are taking place elsewhere. As such, travel itself is seen as a cost. In the context of travel during the course of paid employment, travel time is seen as wasteful – time which could otherwise be put to productive use to the benefit of the employer and economy. Taking the case of individuals travelling on business by rail, this chapter examines travel time use and whether or not the time is indeed wasted. It draws upon a mixed-method research study of 'travel time use in the information age' funded by the UK's Engineering and Physical Sciences Research Study between 2004 and 2007.

The chapter illustrates, with reference to focus group discourses, how the environment of train travel can represent an ideal mobile workspace. Response data from the National Rail Passengers Survey serve to offer a national snapshot of how business rail travellers use and value their time and to highlight therein the place of mobile technologies. An ethnographic account of how rail travellers manage time and space is examined – an approach which reaches beyond what a survey can achieve in connecting with the *experience* of travel. A diary-interview approach explores, at the level of the individual, the place of travel time use in the wider (working) day. The combined approaches reveal travel time use to be a complex interplay between activities and technologies with the boundaries between work and leisure less well defined than the terminology 'business traveller' might initially indicate.

Transport appraisal and the business traveller

Mackie *et al.* (2003) consider two types of business travellers – those whose travel is an integral part of their jobs (e.g. service engineers, delivery people, public transport drivers) and employees travelling in the course of business (termed 'briefcase travellers'). This chapter concerns itself with the latter. However, for both types of business traveller, the same principles are applied when the cost effectiveness of government investment in transport schemes is assessed in transport appraisal. The focus is upon travel time *savings* rather than travel time *use*. Business travel time is seen as time which, if it can be saved (i.e. the journey duration is reduced) will be converted from unproductive time to productive time (Holley *et al.*, forthcoming). This saved productive time is assigned a monetary value which in essence is derived from the average wage rate of business travellers who use a given mode of travel (DfT, 2004). An hour of rail business travel time is costed at £36.96 per individual (ibid.). In contrast, an hour of rail travel time for commuting is costed at only £5.04. All other rail travel outside the course of work is costed at £4.46. These lower values reflect the travel taking place in an individual's own time rather than her employer's. The values are based on averages of individuals' willingness to pay for saved travel time – values estimated using stated preference surveys.

Thus slower journeys are seen as a greater hindrance to economic productivity than faster journeys and, accordingly, huge investments have been made to speed up travel. Such investments have seemingly ignored the possibility that travel time may not be (as much of) an economic burden as supposed. Although there have been a number of critical examinations of the validity of treating travel time as economically wasteful (see Holley *et al.*, 2008; and, for example, Harrison, 1974; Hensher, 1977; Fowkes, 2001; and Mackie *et al.*, 2003), this core assumption has remained in force in transport appraisal for some 40 years. At the same time, there is an acknowledgement that

> the opportunity to use travel time productively can be expected to impact on the value of time, and in this respect the advent and widespread ownership and use of mobile phones and the possibility to use laptop computers on some modes may have had a significant downward influence on the value of time. Future developments may further increase the quality and quantity of useful activities which can be undertaken whilst travelling.
>
> (Mackie *et al.*, 2003: 50)

Observations such as this act as a prompt for this chapter to examine more closely the use of travel time itself with a view to bringing into question whether transport appraisal should account for this.

Mobile work in a technologically enhanced world

It is evident that business travellers *do* incorporate work activities into their journeys from the limited social science research in this area (Letherby and Reynolds, 2003, 2005; Laurier and Philo, 1998; Perry *et al.*, 2001; O'Hara *et al.*, 2002; Brown and O'Hara, 2003).

Mobile technologies can now assist in reproducing the connections and data sources generally experienced in the office environment, which allow for a suitable degree of flexibility in the organization and timing of activities or work tasks (Holley *et al.*, 2008). Laptops (and other mobile devices such as personal digital assistants (PDAs)) make it possible to carry and access much larger amounts of information and resources than previously possible, and the mobile phone facilitates access to colleagues and information located elsewhere. Thus a relatively confined travelling space can be transformed into an environment more akin to the office (see Laurier and Philo, 1998; Perry *et al.*, 2001).

In the case of the car, the passenger seat becomes a storage and retrieval space for documents and samples, while the mobile phone connects the mobile worker with the office and clients, allowing rearrangement of schedules along the way (Laurier, 2004). Travel time use contexts and the relevance of supporting technology artefacts can vary across different modes and at different points in a journey. Brown and O'Hara (2003) found that, while many travellers carried phones, laptops and PDAs, emails were only read online where reliable connections could be made, and paperwork was often used on trains and planes. The redundancy of mobile technologies, such as on a plane, gave some business travellers time to concentrate on reading lengthy documents that they had saved for the trip without the interruptions that occur in the office. Thus, travel time can also open up opportunities for tasks that may suffer from a time squeeze in the office.

A key consideration emergent from the cited work above is the way in which travel time use is planned and appropriated, whether employing new mobile technologies, or more traditional paper-based means, in relation to a range of tasks that need completing as part of a person's job role. Management of the journey time forms an intrinsic part of overall time management and task allocation, reflecting a pre-industrial taskscape concept rather than a structured notion of time organization (see Ingold, 1995; Adam, 1995). This is reflected in the qualitative research below, and is implied in the later quantitative survey evidence.

The ideal mobile office

Focus group research indicated that train travel was perceived as an ideal environment for working compared to other modes. Six gender-defined focus groups were conducted in London, Bristol and Cumbria. The selection of focus group participants aimed to represent a cross-section of age, social class and travel mode (car and public transport users), and to include some people who take mobile information and communication technologies (ICTs) with them on journeys. Since few of the women recruited travelled by train for business purposes, the narratives presented briefly here have a gender bias. The focus group discussions aimed to capture a wide range of travel experiences and the use of mobile ICTs.

Train travel generated an idealized notion of 'potential to work' based on the journey duration and the expectations of the travel space. A business journey was usually *envisaged* as long (i.e. one hour +), and often to and from London. From participants' experience of inter-city travel, the presence of tables and therefore space to work was assumed with scope for accessing a laptop (in contrast, perceptions of a bus journey aligned with much more restricted space). Discussion nevertheless acknowledged that train journeys could vary – not always affording the availability of a table and, in contrast, sometimes presenting the individual with overcrowding, limited space and activities of other passengers all reducing the potential for work. This said, participants also *compared* the train and office environments with suggestions that the former could be more conducive to effective working than the latter – a travel space of welcome retreat. As Brown and O'Hara (2003) indicate, being away from office interruptions can facilitate time for activities that require concentration, such as reading lengthy documents.

However, the potential to work on business travel does not necessarily lead to 'all work and no play'. Rail travel was seen as an opportunity to intermingle work with rest and leisure (see also Jain and Lyons, 2008):

> JAMIE: I sometimes have to go to head office which is down at London so it's quite a long journey so I kind of do a bit of everything, try to snooze, try to relax, every now and again I try and do a bit of work but I normally steer away from that where possible, unless I'm really busy ... I've just got time for myself and time to catch up work and studying and ... just catch up with friends sometimes as well [by mobile phone] (Cumbria).

Only one (male) participant across the groups indicated any sense of *obligation* to work for the duration of a business journey. It appears that individuals are judging for themselves how to get the most value from the journey time – engaging in and moving between 'tasks' (including the task of seemingly doing nothing) which relate to their well-being and work commitment.

How then might these initial qualitative insights relate to a broader national picture of business rail travel time use?

Surveying Britain's business rail travellers

To address this, the authors had the opportunity to design a set of questions on travel time use for inclusion within the autumn 2004 wave of the longitudinal National Rail Passengers Survey. This survey takes a snapshot of rail travel across Great Britain, focusing upon the specific train journey immediately following the receipt of the paper questionnaire by a prospective respondent. The survey distinguishes between individuals travelling by rail for commuting, business and leisure. The response sample for autumn 2004 was 26,221 with survey results weighted to be representative of national passenger rail. For information on the survey methodology and a more detailed discussion of the findings, see Lyons *et al.* (2007).

Table 6.1 highlights the most prevalent time uses of business travellers and

Table 6.1 Travel time usage statistics (from UK National Rail Passenger Survey, 2004)

Activity	*Spent most time (%)*	*Spent some time (%)*	*I made very worthwhile use of my time (%)*	*I made some use of my time (%)*	*My time was wasted time (%)*
Working/studying	31	51	42	54	2
Reading for leisure	25	47	23	63	12
Window gazing/people watching	13	53	12	58	28
Talking to other passengers	5	13	24	56	19
Sleeping/snoozing	3	13	15	57	27
Text messages/phone calls – work	2	22	39	58	2
Text messages/phone calls – personal	1	15	26	50	12
Eating/drinking	1	21	19	80	1

Notes

Specific activities upon which business travellers spent most or some of their journey time (percentage of business travellers) and the corresponding assessment (percentage of relevant respondents) of the journey time use by those spending most time on a given activity.

Not all activities offered in the survey question are included in the table – only those that were selected by at least 10 per cent of respondents for either most time or some time.

reports subjective assessments of the 'value' of the journey time. The results indicate the diversity of time uses. 'Working/studying' is the activity most *obviously* aligned with economically productive use of employers' time and is the single most prevalent activity upon which most travel time by a business traveller is spent. Correspondingly, almost all individuals engaging in this as their main activity on the journey judge the travel time to have been of at least some use. Intriguingly, the majority of business travellers represented in the table rejected the suggestion that their travel time was wasted with indications instead of some subjective benefit. Indeed, though not shown in the table, only 9 per cent of business travellers spent some time being bored on the journey (with only 1 per cent spending most time being bored). In one sense this may bring into question the assumption made in the economic assessment of travel time that such time is wasted.

However, the assessment of 'worth' of travel time reported is that of the travelling employee and not their employer. More than one in ten business travellers who spent most time window gazing/people watching considered that they had made very worthwhile use of their train journey – would/should their employer agree; and was this worthwhile use in a personal well-being and/or work productivity sense? Hence the question remains – has the time for business travel been unproductive for the employer; or, in the absence of the journey or with a shorter duration journey, would the employee have been more productive overall?

Answering such a question was neither an intention or possibility within the confines of the quantitative survey. However, as findings from other methods reported elsewhere in this chapter suggest, the answer to the question is neither clear or straightforward to obtain. Indeed the ownership of business travel time itself is not clear as might be otherwise suggested by the approach to economic assessment.

What is clear, however, is that for some (at least a substantial minority of) business travellers, the environment of the train carriage is able to be put to good use. Indeed, the vast majority of business travellers considered the train an office on the move – 86 per cent answered 'yes' to the question 'in terms of your paid employment is there some work that could easily be undertaken on the train?'.

The significance of mobile technologies to the time use of rail business travellers was also briefly examined. one in five individuals had a laptop with them and one in ten had a PDA. Meanwhile, three-quarters had a mobile phone. Less than half of individuals with such mobile technologies used them during the journey. Three-quarters of those who had laptops and had used them thought that electronic devices made the travel time a lot better and 87 per cent thought that they made the time seem to pass more quickly. The views were similar for those having and using PDAs but less widespread for mobile phones. From these results it seems that mobile technologies are not (yet) pervasive in travel time use nor always essential accessories for time use for those who have them. However, they can add to the flexibility of how time *can* be used and, when they do align with tasks an individual wishes to perform, it seems that they prove valued.

While this large-scale quantitative survey can go some way to illuminating the heterogeneity of business travel time and indeed highlighting the often 'productive' nature of time use, the results offer, at best, only limited explanatory power. While space for time use questions in this particular survey was a limiting factor, as this chapter seeks to demonstrate, other research methods are (also) necessary to begin to assemble a more complete picture of how and to what end business travel time is used. The next section introduces insights from travel ethnography.

On the edge of the working day

First class on the 06:00 train from Newcastle to Kings Cross (a Wi-Fi-enhanced GNER Mallard train) epitomizes the 'ideal business journey'. There is table space for working; the majority of travellers spend two to three hours on the train; and laptops can connect with the online world. While the spatial configuration of this train is unique to the route, thus potentially providing an experience somewhat different to many standard class journeys, the narrative presented from this research illustrates that travel time use is fluid throughout the journey, responding to contextual relationships of place, time, technology and obligations. The researcher (Jain) travelled five consecutive days as a passenger on this route, while observing the activities of other passengers and staff in the immediate area visible from her seat. The aim of this ethnography was to observe how people order time and space for the duration of the journey, and the interaction between electronic mobile devices and other carried objects such as documents, maps and diagrams. The following paragraphs draw upon Jain's field notes.

First class is serviced by a continuous supply of food and drink brought to the seat. This servicing structures the spatial and the temporal organization of the carriages and the intersection of work and nourishment. Thus, the laying out of personal objects to be used during the journey has to contend with the breakfast layout for table space – constraining the environment for working (see Figure 6.1). Many read *The Times*, which is handed out free, as they wait.

'Brian', a middle-aged man who sits across the aisle from me, refuses the paper and starts the journey working on his laptop. As there are internet pages open on his screen I assume he is using the Wi-Fi connection. He starts with room to spread possessions out across adjacent seats, but is interrupted by breakfast. He moves his computer when his [breakfast] plate arrives. He doesn't quite know what to do with it and then moves the cutlery in the seat opposite and repositions it. A working breakfast!

Arguably, the servicing of space constrains the opportunities to work, but travelling this early consumes personal time. Thus, the servicing of personal time at home is shifted to the train. Breakfast and catching up on the news is relocated in space, if not in time.

As the breakfast plates are cleared and newspapers are put away, work becomes the central focus. As most people travel solo, there is little noise from the passengers. I liken the atmosphere to a library or subdued gentleman's club – most passengers are men. There are few phone calls despite phones being strategically poised on the tables, and calls that occur are reporting into home. The available space for working is still limited, despite being more than 'standard class'. On one journey a man opposite me moves tables – he tells me 'more elbow room' (Figure 6.1).

Most people work with documents. Some carry printed-out emails, which they read and annotate, and occasionally co-workers share and discuss documents and plans. The lowest technology offers reliability, travellers tell me, concurring with the other studies of mobile workers. Perhaps too, documents are more flexible in this space, and the lack of phone calls early in the morning enables greater concentration (see Brown and O'Hara, 2003). A few passengers also have laptops, but not many exploit the Wi-Fi potential.

'Brian' is typical of those travellers negotiating documents and a range of mobile technologies. As the journey progresses, Brian moves between activities, and reorganizes and contracts into his space as others joined his table at the various stops. The laptop is awkward and it moves from the table to balanced between lap and table, and back to the table. The space is seemingly slightly less than the ideal, and productivity seemingly discontinuous as he flits between mobile technologies and gazes out on the world.

Fellow travellers fluctuate between spans of concentrated effort to window

Figure 6.1 'More elbow room' negotiates breakfast crockery.

gazing, watching others, responding to the constant offers of tea and coffee and checking their phones for messages. This restlessness and flitting suggests that even in the most idealized journey space, i.e. first-class rail, assumed sustained concentration is an unnecessary myth. There is a need to make breaks and intermingle business with relaxation and play. Some travellers are obvious in their withdrawal into leisure time with iPods (or similar), DVDs and leisure reading, and only rarely does a passenger spend the entire journey gazing out the window and watching others, although a few do sleep.

As we move towards London and 'office hours', the 'work' calls begin and people orientate themselves into the working day. There are rituals of ordering, sorting, packing and connecting. In the final 15 minutes, as the train moves through the London suburbs, where delays sometimes occur, there is a sense of suspended time – a legitimate time to idle or fritter. Mobile phones are rarely packed away, but are poised just in case, then picked up at the last moment. A few people continue to work for a few minutes after arrival and then quickly pack bags as the train crew clean up around them.

Directly observing and sharing in the environment of others travelling extends an understanding of the earlier 'facts' from the survey data. It enriches an appreciation of how the travel space is managed along with the use of mobile technologies. The organization of space and time is constantly constrained by others – passengers and staff. However, people still manage to work and negotiate the constraints (as they may do in other work environments). Other activities, especially eating, are incorporated into the flow of work, rest and play, indicating that time use is seldom neatly partitioned and that the distinction between productive and unproductive time and those tasks which may contribute to both is not easy to make.

Travel time and the working day

The preceding methods have focused upon the journey time itself. However, it is important to appreciate journey time use (and understand it) in the setting of the travellers' overall working and non-working lives. This has been attempted via another methodology involving case studies of individuals. For each case study an initial in-depth one-on-one interview served to establish contextual information followed by the participant completing a diary record for two full days, one based at the normal place of work and the other involving business travel. This was followed by a further interview to discuss the recorded events and allow the participant to enrich her personal account of time use in the working day.

One such 'case study' is that of Alice, who manages an organization-wide electronic library system. The job role itself mainly involves answering queries by email and can be (and currently is being) done almost exclusively from a laptop at home. From the initial interview and a diary kept on a working day without business travel, it became clear that Alice has attempted to implement a rigid time structure to her working days starting regularly at 8.30 am and finishing at 5 pm with an hour break for lunch. From the diary and subsequent discussion it was then possible to compare this day and its time uses with a day in which she travelled by train from Bristol to Exeter to attend a training session:

ALICE: 7.55 I was on the train, read the paper for five minutes, I looked out the window for a total of ten minutes, and for a total of 55 minutes, aren't I good, I prepared for the day.

RESEARCHER: In terms of when you would have read the paper, would that have been as soon as you got on the train?

ALICE: Probably, first thing I did yeah....

RESEARCHER: And then looking out the window, was that in a ten-minutes chunk?

ALICE: No that wasn't in a block it was just in-between, it was probably whilst I was thinking about things.

RESEARCHER: [...] as soon as you got on the train you read the paper for a bit ...?

ALICE: [...] yes definitely, [...] the reason I did that, that way round, is probably because when I find a seat I want to be sure that it's going to be quiet, that I've got a decent window, that I can look out, and it's only when I'm sure that I'm comfortable that I get all my work documents out.

Each of the case studies yielded a different insight into the unique ways in which people organize and adapt their travel time activities to fit in with their working and non-working lives. However, like Alice, (and possibly as a result of completing the diary) all of the six case studies that involved business travel by rail (other case studies involved business travel by car, as a driver and as a passenger, and by plane) involved conducting work-related activities while travelling.

In common with the literature and focus group findings, the tasks chosen were often ones that it was thought were not possible while 'in the office' due to other demands and distractions. Again this was true of Alice, who read the training manual for the day ahead. This was valuable preparation and allowed her to glean more benefit and enjoyment from the day. Had there not been a need to travel for the training programme, this preparation would not have occurred due to the demands on Alice (in common with most of the other case studies) to constantly be checking, reading and replying to emails. Conversely, this is also something which technological advances are making increasingly easier while travelling, which, on this occasion, would have been to the detriment of Alice's productivity.

Alice's organization of time would also mean that a shortening of the journey would not have reduced what would be referred to in appraisal as 'unproductive time' (i.e. the newspaper reading and window gazing), which was used as a buffer period prior to working.

Alice's return journey furnished a very different picture of business travel time use:

ALICE: Got on the train at twenty-five past five ... oh dear, looks like I wasn't very productive ... I think I was pretty shattered actually, so I listened to my iPod, and read the *Metro* paper ... and the train magazine, that's pretty desperate isn't it, those magazines you get on the train about film stars, I think I must have been really tired.

In contrast to the outward journey, the return journey did not involve any work-related activities. In common with many of the other case studies, with nothing specifically planned, Alice conducted less intensive activities on the journey home (although Alice, and one other participant, maintained a pile of work-related reading material in the office from which items are taken when going on train journeys). In this instance technology was deployed, not for productivity, but to help relax after a busy day, with the iPod providing a way of isolating Alice from the surrounding world in a 'privatised auditory bubble' (Bull, 2005), which reduces the impact of external and (due to tiredness) unwanted, stimuli.

When describing her desperation in reading the train magazine, Alice shows some guilt for not conducting any 'productive' activities – perhaps the wasted time considered by transport appraisal. However, Alice's guilt would appear to be inconsistent with her usual work time organization in the office. She had already achieved a full day's work on this business travel day (in terms of hours) and had had a meeting rather than her normal hour lunch break. Other participants felt that they worked and achieved a sufficient amount at other times to be able to travel for business purposes without any obligation to work, especially if the return journey extended beyond the normal working times. This fits with Alice's experience, where the travel time had not affected the time spent working and a reduction in the travel time would have led to increased time at home conducting household and leisure activities (for which she felt there was insufficient time). It was therefore personal time rather than work time that was being sacrificed (recall the different monetary values of time referred to earlier in the chapter).

By looking at the travel time in the context of the entire day it was possible to see that for this particular day it was the only time without any demands, which may indicate a value even if it was not perceived by Alice. This could be either a value in terms of her personal well-being, providing a break and time to mentally adjust between work and home life; or it could have some productivity benefits, allowing time to process, at least subconsciously, the information that had been collected throughout the day (see Csikszentmihalyi and Sawyer, 1995 and Holley *et al.*, 2008).

Conclusion

Business travellers shape their train journeys around their individual needs and in response to the available space and the technologies and artefacts that accompany them. Travel time has a loose structure around potential activities and tasks. Travellers often 'equip' themselves for a range of task options for a journey, but the actual tasks or series of tasks engaged in and intertwined is determined once the context of the journey is established. In particular, the time of day, direction of travel and the pressing need to complete tasks impact on concepts of 'productivity'. It is apparent across the data generated by the different methodologies, that business travel time is an opportunity for working, but it is not consistently dedicated to work output.

Journey beginnings and endings play a specific role as the ethnography and diary interviews indicate. The short period leading up to and just after departure is the settling in and sorting out time, like the planning for the day ahead in the office. Packing up takes a similar role of preparation for the next stage of the journey. Such time can (now) be given to mobile communication – personal and work-related phone calls, perhaps reflecting an activity that can usefully fill a shorter period.

For many business travellers the outward journey, if not the return, is considered an opportunity to work. Thus, we see travellers coming prepared to do work, and valuing this time use. Yet the selection of tasks can often remain very paper-based, despite the increasing potential of mobile technologies to connect beyond the travel space. Paper remains reliable, and is sometimes more manageable in negotiating table space, especially in constrained circumstances, and, as other research indicates, dedicating journey time to reading tasks can be worthwhile.

Most mobile technologies incorporate the dual role of work and leisure, and the qualitative data indicates that people select DVDs to take on journeys to watch when they do not feel obliged to work. Games are available on most mobile technologies and occasionally entertain the business traveller. The various music systems are more specific in the direct withdrawal from the travel space in order to relax as indicated by the travel ethnographies and diary interviews. The business journey, therefore, is not always dedicated to work and sometimes facilitates leisure or relaxation time. Attempting to evaluate the productivity of travel time is likely to be problematic due to the fluid movement between activities on the move.

The chapter's intention was not to directly challenge the established treatment of travel time in transport studies and specifically transport appraisal. However, the insights it has revealed from a variety of research methods certainly suggest that travel time use has a significance in our lives (working and personal) and is intertwined in such a way so as to bring into question whether such a stark and simplifying treatment of its value (or the value of saving travel time) can be fully justified.

References

Adam, B. (1995). *Timewatch: A Social Analysis of Time*. Cambridge: Polity Press.

Brown, B. and O'Hara, K. (2003). 'Place as a Practical Concern of Mobile Workers', *Environment and Planning A*, 35: 1565–1587.

Bull, M. (2005). 'No Dead Air! The iPod and the Culture of Mobile Listening', *Leisure Studies*, 24: 343–355.

Csikszentmihalyi, M. and Sawyer, K. (1995). 'Creative Insight: The Social Dimension of a Solitary Moment', in R. J. Sternberg and J. E. Davidson (eds) *The Nature of Insight*, Cambridge, MA: MIT Press.

DfT (2004). Transport Analysis Guidance: Introduction to Transport Analysis. www.webtag.org.uk.

Fowkes, A. S. (2001). Principles of Valuing Business Travel Time Savings. ITS WP 562.

Harrison, A. J. (1974). *The Economics of Transport Appraisal*. London: Croom Helm.

Hensher, D. A. (1977). *Value of Business Travel Time*. Oxford: Pergamon Press.

Holley, D., Jain, J. and Lyons, G. (2008) 'Understanding Business Travel Time Use and Its Place in the Working Day', *Time and Society*, 17: 21–46.

Ingold, T. (1995). 'Work, Time and Industry', *Time and Society*, 4: 5–28.

Jain, J. and Lyons, G. (2008). 'The Gift of Travel Time', *Journal of Transport Geography*, 16(2): 81–89.

Laurier, E. (2004). 'Doing Office Work on the Motorway', *Theory, Culture and Society*, 21(4/5): 261–277.

Laurier, E. and Philo, C. (1998). Meet You at Junction 17: A Socio-technical and Spatial Study of the Mobile Office. ESRC Report, Department of Geography, University of Glasgow and ESRC.

Letherby, G. and Reynolds, G. (2003). 'Making Connections: The Relationship between Train Travel and the Process of Work and Leisure', *Sociological Research Online* 8(3).

Letherby, G. and Reynolds, G. (2005). *Train Tracks. Work, Play and Politics on the Railways*. Oxford: Berg.

Lyons, G., Jain, J. and Holley, D. (2007). 'The Use of Travel Time by Rail Passengers in Great Britain', *Transportation Research A*, 41(1): 107–120.

Mackie, P. J., Fowkes, A. S., Wardman, M., Whelan, G., Nellthorp, J. and Bates, J. (2003). *Value of Travel Time Savings in the UK*. Report to Department of Transport.

O'Hara, K., Perry, M., Sellen, A. and Brown, B. (2002). 'Exploring the Relationship between Mobile Phone and Document Activity during Business Travel', in Brown, B. Green, N. and Harper, R. (eds) *Wireless World: Social and Interaction Aspects of the Mobile Age*, London: Springer. pp. 180–194.

Perry, M., O'Hara, K., Sellen, A., Brown, B. and Harper, R. (2001). 'Dealing with Mobility: Understanding Access Anytime, Anywhere', *ACM Transactions on Computer–Human Interactions*, 8(4): 323–347.

Tipping, D. G. (1968). 'Time Savings in Transport Studies', *Economic Journal*, 78(312): 843–854.

7 Geographies of international business travel in the professional service economy

James R. Faulconbridge and Jonathan V. Beaverstock

Introduction

International business travel is now a familiar practice for executives. Evidence of this can be seen in the proliferation of hotel chains claiming to specialize in meeting the business traveller's every need (e.g. Marriott Courtyard), websites and magazines providing up to the minute information about the latest services available to business travellers (e.g. Businesstraveller.com), and business-class-only airlines (e.g. Silver Jet) offering the ultimate mobile office. While accepting that it is often hard to disentangle travel that is solely orientated for business purposes and that which also serves tourist or social (familial) functions (Lassen, 2006), it seems undeniable that the globalization of both service and manufacturing industries has led to increasing obligations of proximity and the need for face-to-face contact (Jones, 2007), what Urry (2003) calls 'meetingness'. With this in mind, it is perhaps surprising how little attention business travel has received from academics. One indicator of this research void is the number of business travel articles identified by the ISI Web of Knowledge. Here, as of June 2007, we find 427 articles, a somewhat lowly figure when compared with studies of other forms of mobility such as tourism (7,638 articles) and pilgrimage (1,876 articles).

But why is researching business travel important? There are, we argue, a number of reasons. First, and most pragmatically, the sheer proliferation of business travel journeys. As firms globalize and technological advances seemingly fail to replace the benefits of face-to-face contact, workforces are becoming ever more mobile. The latest Barclaycard Business Travel Survey (2006) suggests that the number of business travel miles of leading executives increased by 32 per cent to 608.5 miles a month on average between 1996 and 2006, something that is predicted to increase by a further 17 per cent by 2015. This has impacts on the way technologies are deployed to support this hyper-mobile workforce (Forlano; Lowry and Moskos, in this volume); social implications for work–life balance (Middleton, in this volume); environmental impacts as a significant proportion of business travel is international and involves air travel; and implications for future urban planning as the demand for airports, airport hotels and business traveller infrastructures leads to calls for new runways and airport developments. Second, theoretically, the study of business travel as a practice can inform wider debates associated with the mobilities paradigm (Cresswell, 2001; Sheller and Urry, 2006), global sociologies and microglobalization (Knorr-Cetina, 2006; Urry, 2000) and relational economic spaces and geographies (Amin, 2002; Yeung, 2005). We need to move beyond studies of the functionality of business travel in the sense of 'getting the job done' and consider the new relational geometries, 'network capital' (Larsen *et al.*, 2006) and topologies of business space that are opened up by travel. These have important geographical dimensions that reinforce forms of uneven development in the global economy.

With this in mind, the rest of the chapter focuses exclusively upon the international dimensions of business travel in three main ways. First, we illustrate the contemporary patterns of business travel through an analysis of official data from the United Kingdom's (UK) *Travel Trends* (Office for National Statistics, 2006) and unofficial data, sourced from the European airline industry (Witlox *et al.*, 2007). Then we examine the way this proliferation in business travel supports the operation of globalizing professional service firms. Drawing on examples of how lawyers use business travel, we show the topologies of social space opened up by occasional face-to-face encounter and explain how these serve to sustain working relationships and facilitate transnational business. We conclude by raising a number of important questions about the way geographies of business travel reproduce the global space economy.

Overview: mapping the business travel boom

Recording the flows and magnitude of international business travel in the world is not an exact science. Data are generated from either national governments or private enterprise for commercial purposes (e.g. airlines and hotel chains to predict passenger yields and bed occupancy).

Official business travel trends

In the UK for example, the Office for National Statistics (ONS) collects data on overseas residents' visits to the UK and UK residents' visits abroad in the annual statistical digest, *Travel Trends*. Data reported in *Travel Trends* are collected from the International Passenger Survey, a continuous face-to-face interview survey of a sample of passengers entering and leaving the UK by sea, air and tunnel routes (which reports the purpose of visits as: holiday; business; visiting friends or relatives; and miscellaneous) (ONS, 2006). In the following, we present the major travel trends in *business visits* to and from the UK as recorded in the most recent *Travel Trends 2005* (ONS, 2006):

- In 2005, 8.556 million UK residents made visits abroad and 8.168 million overseas residents entered the UK (Table 7.1);
- the annual rate of growth of business visits for both UK residents leaving

and overseas residents entering the UK is approximately 5.1 per cent (Table 7.1);

- the average length of stay of UK residents leaving the UK is 5.4 nights and overseas residents staying in the UK is 4.2 nights between 2001 and 2005 (see Tables 3.05 and 2.05, ONS, 2006);
- UK residents' visits abroad and overseas residents' visits to the UK for business increased by 169 and 171 per cent respectively between 1985 and 2005 (Table 7.1);
- the share of business visits as a percentage of all visits has always been greater for overseas residents entering the UK than UK visits abroad, which average at 25 and 15 per cent respectively between 1985 and 2005 (Table 7.1);
- Europe is by far the most important region for residence of overseas visitors entering and UK residents leaving the UK, accounting on average for 78 and 82 per cent of all business visits respectively between 2001 and 2005 (Table 7.2);
- in 2005, a quarter of all overseas business visitors entering the UK had their country of residence in either Germany (13 per cent) or France (12 per cent) (Table 7.3), and 39 per cent of all overseas business visitors specifically visited London (Table 7.4);
- in 2005, over a third of all UK residents' business visits abroad were to France (15 per cent), Germany (11 per cent) and the Republic of Ireland (10 per cent) (Table 7.3).

Accordingly, this brief foray into the UK official data shows very clearly that business travel is not only increasing in absolute terms year on year for both UK and overseas residents leaving and entering the UK, but also that it remains a vital business process irrespective of distance travelled. Given all the rhetoric about ICT, video-conferencing for example reducing the requirements for physical travel in modern-day business life, perhaps the last region that one would have expected to be the most important location for destination and origin of business travellers is Europe, and the UK's nearest neighbours, Germany, France and Ireland! It is also very significant to note that London's corporate economy is at the hub of business travel into and out of the UK.

Unofficial business travel trends

One very rich seam of unofficial data on international business travel can be procured from the airline industry. While such airline data is highly commercially sensitive and confidential when disaggregated into different classes of travel (first, business, premium and economy), two world experts on airline travel flows, Frank Witlox and Ben Derudder, have been setting the international agendas in mapping passenger travel data for the airline industry (e.g. Derudder, Devriendt and Wiltox, 2007; Derudder, Wiltox and Taylor, 2007). Witlox *et al.* (2007) uses data published by the Association of European Airlines for the

Table 7.1 Business visitor trends ('000s), 1985–2005

	UK residents' visits abroad		*Overseas residents' visits to the UK*	
	Visits	*% of total*[1]	*Visits*	*% of total*[1]
1985	3,188	15	3,014	21
1990	4,769	15	4,461	25
1995	6,113	15	5,763	24
2000	8,872	16	7,322	29
2005	8,556	13	8,168	27
Annual growth 5.1%			Annual growth 5.1%	

Source: ONS, 2006; adapted from Tables 1.03, p. 26 and 1.04, p. 27.

Note
1 Total visits include: holiday, business, visiting friends or relatives, and miscellaneous.

Table 7.2 Number of business visits by destination and region of residence ('000s), 2001–5

Overseas residents' business visits to the UK by region of residence						
	2001	*2002*	*2003*	*2004*	*2005*	*Growth 2001–5 (%)*
North America	847	880	803	811	898	1.5
Europe	5,198	5,551	5,500	5,898	6,464	5.6
Other countries	734	728	664	760	806	2.4
Total world	6,778	7,158	6,967	7,470	8,168	10.2
UK residents' business visits by region of visit						
	2001	*2002*	*2003*	*2004*	*2005*	*Growth 2001–5 (%)*
North America	834	727	720	791	855	0.6
Europe	6,731	6,663	6,488	6,588	6,782	0.2
Other countries	655	683	683	761	919	8.8
Total world	8,220	8,073	7,982	8,140	8,556	1.0

Source: ONS, 2006, adapted from Tables 2.03, p. 36 and 3.03, p. 58.

period January 2002 to December 2005 to chart the geography of business-class travel in Europe by converting airport-to-airport-by-carrier data into a city-by-city database by aggregating for business class the number of passengers for all airports of a given city (i.e. includes flights within Europe, and between Europe and other world regions). Table 7.5 shows two very important rankings in Europe's world city geography of airline business-class travel. First, it illustrates the most *absolute* important cities by passenger volume in business-class travel, with London, Frankfurt and Paris atop this hierarchy. Second, and very importantly, it depicts the *relative* proportion of business class travel within a city's

Table 7.3 Number of business visits by country of residence and country of visit ('000s), 2005

Overseas residents' visits to the UK by country of residence			*UK residents' visits by country of visit*		
Residents	*Visits*	*% of total*	*Country*	*Visits*	*% of total*
Germany	1,077	13	France	1,267	15
France	997	12	Germany	971	11
USA	801	10	Eire	862	10
Netherlands	645	8	USA	794	9
EIRE	497	6	Netherlands	641	7
Spain	431	5	Belgium	503	6
Belgium	401	5	Spain	460	5
Poland	391	5	Italy	437	5
Italy	346	4	Switzerland	284	3
Switzerland	218	3	Sweden	190	2
Other	3,364	29	Other	2,147	25

Source: ONS, 2006, adapted from Tables 4.04, p. 82 and 5.04, p. 105.

Table 7.4 Overseas residents' visits to the UK by area of visit ('000s), 2005

Country of visit	*Visits*	*% of all visits*
England	6,465	79
Scotland	401	5
Wales	181	2
Other	1,121	14
Area of visit		
London	3,175	39
West Midlands	474	6
Manchester	368	5
Kent	203	3
Berkshire	195	2
Cambridgeshire	170	2
East Sussex	166	2
Surrey	164	2
Oxfordshire	162	2
Lothian (Scotland)	162	2
Greater Glasgow	158	2
Tyne and Wear	130	2
Bristol	129	2
Hampshire	128	2
Merseyside	127	2
Other	2,257	25

Source: ONS, 2006, adapted from Table 4.11, pp. 94–95.

Table 7.5 Airline business class travel flows in Europe, 2002–5

Rank	*City*	*Business class travel flows*	*Rank*	*City*	*Proportion of business class travel flows*
1	London	3,281,117	1	Geneva	16.26
2	Frankfurt	2,026,604	2	Oslo	16.00
3	Paris	2,019,845	3	Stockholm	14.78
4	Amsterdam	1,737,635	4	Düsseldorf	14.62
5	Copenhagen	1,096,543	5	Frankfurt	13.61
6	Munich	1,080,402	6	London	12.95
7	Stockholm	863,045	7	Zurich	12.77
8	Milan	853,438	8	Copenhagen	11.80
9	Vienna	764,851	9	Munich	11.75
10	Brussels	763,111	10	Vienna	10.75
11	Madrid	759,496	11	Brussels	10.72
12	Oslo	694,255	12	Amsterdam	10.10
13	Geneva	568,867	13	Paris	9.67
14	Rome	514,360	14	Milan	9.32
15	Düsseldorf	461,820	15	Berlin	8.39

Source: Wiltox *et al.* (2007).

overall passenger volumes, which shows that cities like Geneva and Frankfurt are highly influential business centres in Europe.

Given the dearth of both official and unofficial data on business travel in Europe and beyond, a particularly fertile approach to unpacking the processes and patterns of such travel and working practices has been detailed case-study research, often involving qualitative-based face-to-face interviewing. In the following section we use the example of business travel in professional services to illustrate why such working processes are still vital working practices in an era of highly complex ICT as the speed of globalization continues to shrink the business world in an ever-increasing web of interconnections.

Business travel and professional service firms: stretched social relations and topologies of economic space

For Sassen (2006), it is professional service firms (Table 7.6) in particular that underlay contemporary economic globalization. Their activities, mediated in and through world cities, have created global office networks capable of meeting the accounting, advertising, financial and legal needs of other globalizing firms. Significantly, the activities of these firms are, according to Sassen and others (e.g. Beaverstock, 2004; Faulconbridge, 2007a) sustained by a myriad of flows of information, instructions, ideas, plans and importantly here people. It is the peculiarities of professional services and the delivery of integrated global professional services that create this situation. In particular, it is possible to identify two unique characteristics of the work of these firms that render business travel especially important. First, they deliver knowledge-rich, tailored advice to clients (Alvesson, 2001; Lowendahl, 2005). This often requires face-to-face contact between advisors and clients. Second, deals/projects often stretch across borders and bring together professionals (both on client and service-firm sides) from multiple countries. Coordinating and synchronizing this process, as well as developing the trust and mutual respect needed to allow collaboration, often requires intermittent periods of co-presence. The comments of the UK Chairman and Senior Partner of KPMG the accounting firm exemplify these points. He noted that, 'half of our carbon footprint is now accounted for by air travel and we can't quite see how we can deliver services to our international clients without it' (*Financial Times*, 2007).

Jones (2007) develops this idea in his explanation of the role of face-to-face contact in globalizing law firms being facilitated by business travel, listing five reasons:

1 The operation of the firm – securing new business and the completion of projects.
2 The control of the firm – coming to agreement with semi-autonomous partners in different offices about the implementation of strategy.
3 Knowledge practices – producing, sharing and deploying the knowledge of employees.
4 Innovation – the production of new products/services.
5 Coherence – creating a shared organizational culture.

Here, however, we aim to push this debate in a different direction. Rather than focusing on business travel as a functional 'means to an end' (e.g. meeting up to

Table 7.6 Exemplary professional service firms

Industry	*Exemplary firms*	*Country of origin*	*No. offices (countries served for accountancy)*
Accountancy	Deloitte Touche	UK	150
	KPMG	Merger between UK/US firms	148
	Ernst Young	Merger between UK/US firms	140
Advertising	WPP	UK	1,300
	Interpublic	USA	671
	Havas	Spain	340
Architecture	Gensler	USA	26
	RTKL	USA	11
	NBBJ	USA	8
Law	Baker and McKenzie	USA	70
	Clifford Chance	UK	28
	Allen and Overy	UK	25

Source: Firms' websites.

sign a deal), we explore the practice's social significance and the way studying business travel can help us understand the linkages between fundamental sociological ideas of social proximity, face-to-face contact and localness. We build on the arguments of both Urry (2000) and Knorr-Cetina (2006) to examine the way business travel is producing stretched social formations that pull particular places and spaces together into topological assemblages formed through mobility and occasional moments of proximity. Such co-presence acts as a socially aligning and synchronizing force that, as Sheller and Urry (2006) argue, produces complex and fluid social topologies that we know little about at present. Of course, as Callon and Law (2004) point out, we should not only study mobility, but also the various other technologies (the telephone, internet, BlackBerry) that are folded into the construction of this new form of proximity. Here the idea of an encounter moves further beyond the face-to-face and recognizes the multiple ways in which 'being there' can exist. This means the synergism of occasional meetings facilitated by business travel with virtual communications should be the focus of our study so as to develop an understanding of how formal, social, event- and place-based obligations of meetingness exist alongside occasions when virtual co-presence is adequate (Urry, 2003).

From a geographer's perspective, this means recognizing the implications of contemporary forms of mobility for studies of the relational economy (Boggs and Rantisi, 2003; Yeung, 2005). Here attempts to analyse the geographically distributed nature of economic activities have begun to explore the ways that Hess (2004: 180) calls network embeddedness ('the connections between heterogeneous actors, regardless of their location, rather than restricted to only one geographical scale') and Amin (2002: 389) refers to as relational space/proximity ('space, place and time … as: co-constituted, folded together, produced through practices, situated, multiple, and mobile'). Next, we begin to consider this issue and look specifically at the role of business travel as a creator of new topologies of social, economic space in legal professional service firms where both embodied and virtual co-presence ties the firm together. We do this by drawing on insights gained from 29 semi-structured interviews held with managing partners, partners and associates working for globalizing law firms completed in London and New York between 2003 and 2004.

Unpacking business travelling in professional service firms

In some ways it is impossible to divorce the role of business travel from 'getting things' done (Jones, 2007). For the most part, the travel of the lawyers interviewed was restricted to the necessities of a transaction. To adopt the language of Urry (2003), this would normally be for legal, economic and object obligations (the signing of contracts), but also social obligations (meeting the clients at the start and at critical moments in a transaction). At the time of completing the interviews, the financial costs of overseas travel and accommodation, and the add-on cost of the unproductive periods associated with business travel rendered all, but the most important journeys economically unviable. Nevertheless, despite these costs, lawyers were unanimous in calling business travel an essential practice. This was not just because of its relationship with ‘getting the job done,’ but also because of the ‘network capital’ (Larsen *et al.*, 2006) and relational proximity (Amin, 2002; Faulconbridge, 2007b) it produces as new social spaces are opened up between colleagues in different offices. As one lawyer put it:

> if you don’t trust each other you’re not going to work together and share knowledge effectively, and so it’s really finding ways to build trust between groups of people. Now what we try and do is at least at a partner level to try and find ways of building this through partner retreats where people get to know one another.
>
> (L2)

As the quotation suggests, developing these spaces requires the manufacturing of opportunities to develop relationships and connections between individuals, something that can mean staying on an extra day after the completion of a deal or ensuring social time (evening meals, a round of golf) is incorporated into schedules. Indeed, such ‘getting-to-know-you’ time is valued so much by lawyers that it can, on occasions, serve as the sole justification for travel. As one partner commented:

> our team is pretty good at trying to meet people, get out and about. Three of us went to [place x] last week just to meet them, it’s not vital, but it’s massively important. My point of view is it’s brilliant because now if I’ve got a problem, I’ve pretty much met everybody from that office, and they’ll be no problem, I’ll know exactly who to call.
>
> (L17)

Such trips occurred most commonly in the form of annual all-partner conferences or practice group meetings. All agreed that the main value was the networking facilitated as the meetings acted as a mechanism to ‘gel’ the network together and move away from seeing overseas offices as pins in a map and towards a vision of them as part of a shared organizational social space. The following quote illustrates the importance of this aspect of business travel:

> We have been putting more and more emphasis on periodically getting people together and it’s an important thing.... It’s difficult to put your finger on. Just being able to attach a face to a name and remember that you had dinner with that person and talked to them about something can be an important aspect of making you feel personally comfortable with them. Also, just seeing people in those kinds of situations and hearing them talk about the way they approach things and getting a feeling for who they are, it’s very intangible but it gives you a feeling of comfort in how you deal with them ... there’s an atmospheric sense of feeling you know somebody

> that comes from actually spending time in their physical presence not just talking with them.
>
> (L25)

As this suggests, obligations of meetingness can actually transform organizational spaces. Orlikowski (2002) came to the same conclusion when she studied work practices in a multi-site software company. She noted that face-to-face interactions facilitated by business travel led to relations of trust, respect and credibility developing between employees working at different sites. This then allowed information-sharing and sustained social networking that meant knowledges could be produced and reproduced by teams formed of spatially distributed individuals.

The resultant fluid social topologies facilitated by business travel bring into question, then, ideas such as near and far, proximate and distant, as occasional embodied encounters open up new connections across space that drag colleagues together into alignment and topological proximity. Significantly, mirroring the arguments of Callon and Law (2004), interviews revealed that technology plays a supporting role in this process of folding distanciated offices together. Many of the relationships consummated through physical co-presence had been established virtually, usually by email or telephone (see Forlano, in this volume). This developed the initial foundations of a relationship (awareness of one another's expertise and experience) that were then built upon when face-to-face encounters occurred. After the meeting and construction of the main components of a successful relationship (mutual understanding and trust in particular, as detailed by Goffman, 1967) virtual co-presence through regular email dialogue, telephone calls normally at intervals of once a week (see Licoppe, 2004), and occasionally video-conference maintained and sustained the relationship.

The importance of this blending of occasional face-to-face and ongoing virtual encounter and co-presence was clear to all of the lawyers interviewed. Together, the two ensured it was possible to have rich and meaningful exchanges on a regular basis, not just when it was possible to get together in the same room. This is vital to globalizing firms because, as one lawyer put it:

> in the ideal world it would always be nice to meet face-to-face and to be able to work together ... but the reality is that it costs money because someone's going to have to get on a plane.... That's the ideal scenario but in reality people are often going to try and avoid doing that so you'll use the other most effective ways so inevitably you will talk to each other on the phone.
>
> (L3)

Often these conversations serve to deal with all, but the final stages of a transaction and reach agreement about how to structure the legal documents being put together for a client. They may even be used for firefighting when a crisis arises, but time or other business commitments prevent travel for a face-to-face meeting. At the opposite end of the spectrum they are also used as a way of innovating and managing knowledge in the firm (Faulconbridge, 2007a, 2007b). Because of the tacit, embodied nature of the knowledge employed to produce bespoke solutions to clients' problems, the best way to efficiently and effectively provide advice is by seeking support, suggestions and help from colleagues with experience of the type of problems encountered in a transaction. Often these individuals will be both metrically close (in the same office) *and* distant (in overseas offices). As one lawyer put it:

> So when you're up against a problem first you walk down the corridor and talk to your colleagues, but if they can't find a solution and you think there is more mileage to be had out of this you pick up the phone and talk to the partners who you think might have something to offer here, and they might be in Germany, in New York, or in France.
>
> (L9)

But geography matters in business travel!

It became clear that the geography of business travel was a proxy for the intensities of flow and interconnection between places and, most importantly, a proxy for the uneven geographies of the global space economy. As Sheppard (2002) has shown, the geography of the space economy is historically determined, but also reinforced by contemporary practices that open up 'worm holes' that act as short-cuts for economic flows. As he argued, time cannot totally trump space and there is a need to recognize 'relational inequalities within, networked spaces' (Sheppard, 2002: 308). He goes on to 'advance the idea of positionality as a way of capturing the shifting, asymmetric, and path-dependent ways in which the futures of places depend on their interdependencies with other places'. Business travel, because of the social space it constructs, is a major producer of the worm holes Sheppard described and reinforces familiar, global triad-based uneven geographies of economic activity (see Dicken, 2007). This means understanding both the significance of presence in the networks, but also the way a place is positioned in these networks, what role it has and what type of power relations exist between it and other nodes in the network.

There isn't the scope to cover the latter here. Instead we focus upon the former and the way business travel allows some metrically distant places to become closely connected through topological social spaces and 'worm holes' while, at the same time, reinforcing the isolation of other places. In line with the ideas of Nowicka (2006), we argue that, while it is tempting to get carried away with discussions of time-space compression and to completely switch focus onto social topologies of space where metric locatedness does not matter, studying business travel reveals the continued importance of location in relational networks. Interviews revealed that the intensity with which a place was connected to other spaces of economic activity by business travel, i.e. the frequency with which lawyers visited or travelled from an office, continues to be influenced by two factors. First, the 'positionality' of a place in the global economy (i.e. the relative importance of a place in a global hierarchy of commercial centres (Faulconbridge, 2004)) – at its simplest this manifests itself in the presence/absence of transnational law firms. As Beaverstock *et al.* (1999) noted, the geography of the offices of these firms is far from global and is, instead, geographically selective with activities focused, in particular, upon North America and Western Europe with a growing presence in Asia. This geography has been influenced by a globalization process determined, in the case of English-originating law firms, by an imperial logic with offices opening wherever English firms are most active. Meanwhile, American firms have only opened offices where there is significant demand for advice on American law (Morgan and Quack, 2005). Consequently, certain places (countries throughout Africa, Russia to a large degree) continue to be excluded from the new economic topologies of capitalism.

Second, metric forms of space (i.e. distance in kilometres between places) also determine whether offices are included in relational social spaces. With the demise of Concorde, air travel has actually become slower in recent years and those places located both metrically and temporally distant from the North American and English heartlands of globalizing law firms are often excluded (Australia, where none of the leading firms have offices) or suffer from weaker incorporation (much of Asia) into new topological social spaces. Zook and Brunn (2006) show how air travel time maps are both, topological depending on service levels, but also metric because of the continued friction of distance. To travel to/from Asia or Australia from the heartlands of globalizing law firms requires journeys of at least 12 hours and up to 16 hours and the associated costs of at least £3,000 (US$6,000) for a business-class ticket. In comparison, the transatlantic hop between New York and Western Europe or London and the rest of Europe is, both financially and temporally, less costly. This results in less frequent travel to/from offices in such metrically distanciated locations and relative isolation because of limited opportunities to construct the types of social spaces described above. As one lawyer put it: 'our Asia offices, because they are a long way away, have a higher degree of autonomy. Notwithstanding the developments in information technology and communications, they are a long way away' (L2). The interviewee sets up the situation as potentially beneficial, suggesting that the autonomy that metric distance grants offices is liked by lawyers in Asia. However, if we consider all of the advantages associated with relational, topological social spaces laid out above, this seems more problematic. Of course, these situations are constantly in flux and, as countries in Asia become more and more important actors in the global economy, this situation might begin to evolve. In this sense, as Callon and Law (2004) argue, the topologies of the global space economy and the social spaces associated with its enactment are perhaps best conceived as a gel, multiple, open and constantly changing in fluid ways as new contexts and priorities reshape their geographies. Nevertheless, as this example shows, we should not assume that this displaces the need for an understanding of the very real impacts of positionality and metric locatedness in these new forms of mobility and associated connectivity.

Conclusion

We began by suggesting that the growing importance of business travel in the lives of executives made it surprising that the topic had received such limited attention. For scholars of geography, management and sociology, there seem to be many unanswered questions about the role, effects and ways of managing business travel in the contemporary economy. In this chapter we have begun to point towards some of what we see as the most interesting questions in this respect. We have called for studies to move beyond examining solely the functionality of business travel and promoted analyses that seek to understand the role of business travel in reproducing the global space economy. In this respect, two conclusions are particularly significant.

First, the proliferation of business travel journeys and the seemingly dynamic nature of their geographies make understanding the added-value gained from face-to-face encounter and meetingness essential (Orlikowski, 2002; Urry, 2003). We have begun to flesh out a number of important ideas about how business travel and embodied encounters enable, alongside virtual encounters, the construction of new forms of 'network capital' (Larsen *et al.*, 2006), relational proximity (Amin, 2002; Faulconbridge, 2007a) and network embeddedness (Hess, 2004). Most importantly this reinforces recent suggestions (Urry, 2000, 2003; Sheller and Urry, 2006) that fundamental sociological ideas such as trust and familiarity cannot, due to mobility trends in the current epoch, be solely associated with metric closeness. This is not to say that embodied encounters are not important. Instead, as the examples of activities in globalizing law firms shows, it is important to recognize the new strategies being employed by firms to open up the types of social spaces needed for effective business.

This discussion does, however, raise some important questions. For example, we know little about how frequent repeat travel needs to be after an initial embodied encounter to prevent the decay of any relationship established. In the language of Callon and Law (2004), it seems that such relationships may be flickering and in need of long-term nurturing. At present we don't fully understand the temporal dynamics of this. Similarly, we don't know enough about how the duration and nature of an embodied encounter might influence the construction of topological social spaces. Licoppe (2004) raises a number of interesting issues when he describes how phone calls occur more or less frequently and for shorter or longer durations depending on geographical proximity. It would seem important, then, to consider if the variations in the duration of embodied encounters strengthen relationships and whether the nature of activities (work versus social, etc.) further influences this. Finally, in the context of professional service firms, we do not know what differences exist between intra-firm (colleague-to-colleague) versus extra-firm (professional-to-client; professional-to-strategic partner) relationships and whether business travel can construct the same types of relational space in all cases. Here we have focused upon intra-firm dynamics. Future research should be expanded to also cover the extra-firm dimension.

Second, our findings suggest that we should not write out geography from these discussions. While it is clear that mobility and time-space compression have significantly altered the relationship between metric distance and concepts of proximity, geography still seems to matter in a number of ways (Sheppard, 2002; Nowicka, 2006). As we described in relation to globalizing law firms, this can be in relation to presence/absence and positionality in networks as well as actual metric distance when the frictions of time continue to trump space as a defining feature of social action. This clearly deserves further investigation as understanding the way these influences upon business travel (re)produce the geography of the global space economy is significant. For example, it would seem important to consider how the growing influence of Asian economies on the activities of all global firms is or is not leading to economic priorities overcoming the hindrance of time described above. Moreover, and unexamined here, it would seem important to consider if business travel is used to construct relationships with the same or different managerial and sociological characteristics across space. So, for example, is collaboration always the desired outcome of such relationships or is command and control sometimes more important? How does geography determine this – i.e. are relations between the London and New York offices and the London and Kiev offices of globalizing law firms the same? Our initial investigations suggest not (Beaverstock *et al.*, 2000; Faulconbridge, 2007a, 2007b), but this deserves further attention.

It would seem, then, that there remains a number of important questions about business travel that need exploring. In the future this is likely to become of significance in policy debates (see Wheatley, Hardill and Green, in this volume), not least because of the need to tackle the climate change impacts associated with air travel as business organizations of all sizes and nationalities come under greater scrutiny to reduce their global carbon footprint in all future business activities.

References

Alvesson, M. (2001) 'Knowledge Work: Ambiguity, Image and Identity', *Human Relations*, 54: 863–886.

Amin, A. (2002) 'Spatialities of Globalization', *Environment and Planning A*, 34: 385–399.

Barclaycard Business Travel Survey (2006) Available from www.barclaycardbusiness.co.uk/information_zone/travel/index.html [accessed 07/06/07].

Beaverstock, J. V. (2004) 'Managing across Borders: Knowledge Management and Expatriation in Professional Legal Service Firms', *Journal of Economic Geography*, 4: 157–179.

Beaverstock, J. V., Smith, R. and Taylor, P. J. (1999) 'The Long Arm of the Law: London's Law Firms in a Globalising World Economy', *Environment and Planning A*, 13: 1857–1876.

Beaverstock, J. V., Smith, R. and Taylor, P. J. (2000) 'World-city Network: A New Metageography', *Annals of the Association of American Geographers*, 90: 1, 123–134.

Boggs, J. S. and Rantisi, N. (2003) 'The "Relational Turn" in Economic Geography', *Journal of Economic Geography*, 3: 109–116.

Callon, M. and Law, J. (2004) 'Guest Editorial', *Environment and Planning D*, 22: 3–11.

Cresswell, T. (2001)' The Production of Mobilities', *New Formations*, 43: 11–25.

Derudder, B., Devriendt, L. and Witlox, F. (2007) 'Flying Where You Don't Want to Go: An Empirical Analysis of Hubs in the Global Airline Network', *Tijdschrift voor Economische en Sociale Geografie*, 98: 307–324.

Derudder, B., Witlox, F. and Taylor, P. J. (2007) 'United States Cities in the World City Network. Comparing Their Positions Using Global Origins and Destinations of Airline Passengers', *Urban Geography*, 28: 74–91.

Dicken, P. (2007) *Global Shift*. London: Sage.

Faulconbridge, J. R. (2004) 'London and Frankfurt in Europe's Evolving Financial Centre Network', *Area*, 36: 235–244.

Faulconbridge, J. R. (2007a) 'Relational Spaces of Knowledge Production in Transnational Law Firms', *Geoforum*, 38: 925–940.

Faulconbridge, J. R. (2007b) 'London and New York's Advertising and Law Clusters and Their Networks of Learning: Relational Methodologies and Scalar Reflections', *Urban Studies*, 44: 1635–1656.

Financial Times (2007) 'Air Travel Is the Biggest Challenge', *Financial Times*, 5 June 2007.

Goffman, E. (1967) *Interaction Ritual: Essays on Face to Face Behaviour*. New York: Doubleday and Company.

Hess, M. (2004) 'Spatial Relationships? Towards a Reconceptualization of Embeddedness', *Progress in Human Geography*, 28: 165–186.

Jones, A. (2007) 'More than 'Managing across Borders?' the Complex Role of Face-to-Face Interaction in Globalizing Law Firms', *Journal of Economic Geography*, 7: 223–246.

Knorr-Cetina, K. (2006) 'Microglobalization' in I. Rossi (eds) *Frontiers of Globalization Research*, New York: Springer.

Larsen, J., Axhausen, K. W. and Urry, J. (2006) 'Geographies of Social Networks: Meetings, Travel and Communications', *Mobilities*, 1: 261–283.

Lassen, C. (2006) 'Aeromobility and Work', *Environment and Planning A*, 38: 301–312.

Licoppe, C. (2004) 'Connected' Presence: The Emergence of a New Repertoire for Managing Social Relationships in a Changing Communication Technoscape', *Environment and Planning D*, 22: 135–156.

Lowendahl, B. (2005) *Strategic Management of Professional Service Firms* (3rd edition). Copenhagen: Copenhagen Business School.

Morgan, G. and Quack, S. (2005) 'Institutional Legacies and Firm Dynamics: The Growth and Internationalization of UK and German Law Firms', *Organization Studies*, 26: 1765–1785.

Nowicka, M. (2006) *Transnational Professionals and Their Cosmopolitan Universes*. Frankfurt and New York: Camous Verlag.

Office for National Statistics (2006) *Travel Trends 2005*. London: Palgrave Macmillan.

Orlikowski, W. (2002) 'Knowing in Practice: Enacting a Collective Capability in Distributed Organizing', *Organization Science*, 13: 249–273.

Sassen, S. (2006) *Cities in a World Economy* (3rd edition). London: Sage.

Sheller, M. and Urry, J. (2006) 'The New Mobilities Paradigm', *Environment and Planning A*, 38: 207–226.

Sheppard, E. (2002) 'The Spaces and Times of Globalization: Place, Scale, Networks and Positionality', *Economic Geography*, 78: 307–330.

Urry, J. (2000) *Sociology beyond Societies: Mobilities for the Twenty-first Century*. London: Routledge.

Urry, J. (2003) 'Social Networks, Travel and Talk', *British Journal of Sociology*, 54: 155–175.

Wiltox, F., Derudder, B., Faulconbridge, J. and Beaverstock, J. (2007) 'Airline Business Travel Flows in the Global Space Economy. Quantitative and Qualitative Approaches to Analysis', paper presented at the Association of American Geographers Annual Meeting, San Francisco, California, 17–21 April (copy available from the lead author).

Yeung, H. W.-C. (2005) 'Rethinking Relational Economic Geography', *Transactions of the Institute of British Geographers NS*, 30: 37–51.

Zook, M. A. and Brunn, S. D. (2006) 'From Podes to Antipodes: Positionalities and Global Airline Geographies', *Annals of the Association of American Geographers*, 96: 471–490.

Part III

Mobile work practices

8 The lonely life of the mobile engineer?

Carolyn Axtell and Donald Hislop

While mobile teleworking might be considered a neglected subject in itself (Hislop and Axtell, 2007), when it comes to studies of mobile workers (or teleworkers more generally) many of these focus on professional or administrative roles, and jobs such as engineering have received relatively little attention (Churchill and Munro, 2001). However, mobile service and repair engineers are an interesting group to examine because they might be considered an almost 'pure' example of a mobile worker. This is because they predominantly work at locations beyond their home and employer's office; spending most of their time out in their vehicles or at customer sites (see Hislop and Axtell, 2007). Typically, mobile engineers do not spend much time at their employer's premises, as their main role is to service or repair equipment located with the customer.

This lack of physical presence at the main office is something that mobile work has in common with home-based work. Certainly, in the teleworking literature, being located away from the main office and working on one's own at home has been associated with feelings of isolation (e.g., Daniels, 2000; Haws, 1984; Mann and Holdsworth, 2003). Moreover, a previous study of mobile work found isolation from co-workers to be an issue (Laurier and Philo, 2003). However, the factors that contribute to relieving this isolation have not been examined in detail, and nor has there been any explicit comparison of different mobile work arrangements. Moreover, the role of communications technologies on isolation and relationships more generally has mostly been considered within the 'virtual teams' rather than the (mobile) teleworking literature. Thus linking these two literatures might be helpful in our understanding of these issues.

Both physical and virtual factors may be at play here. First, the level of isolation felt may well depend on the extent to which physical presence at a communal office is achieved. For instance, the mobile workers studied by Brown and O'Hara (2003) visited their main office regularly even if there was no specific reason to do so, so that they would be available for interaction with others. Such interaction enables mobile workers to maintain their professional network and exchange knowledge. Consistent with this idea, studies have also found that part-time teleworking (not permanently away from the main office, but working at home for only a portion of the week) does not result in teleworkers being left out of the office network or disrupt communication patterns (e.g., Belanger,

1999). Thus, being absent on only a part-time basis from the main office appears to be less isolating.

With regards to mobile engineers, there may be some variations in the level of this absence. For instance, some may have a company base that they visit fairly frequently to pick up parts, chat informally to colleagues, attend formal meetings etc., whereas other engineers may not have such a base, instead having parts delivered closer to their home and being obliged to schedule meetings with colleagues at mutually convenient places. It might be predicted, therefore, that those who have a company base, and therefore more frequent face-to-face contact with colleagues, will feel less isolated than those who lack such opportunities.

Second, another predictor of isolation is likely to be the use of technology for 'virtual' interaction. For instance, messages sent via text-based media (such as email) contain fewer social cues and tend to be much shorter and less 'chatty' than telephone conversations (Sproull and Kiesler, 1986). Text-based media, especially when used asynchronously, might therefore be considered to be 'low in richness' as a result of the lack of auditory and visual cues they can transmit and the lack of immediate feedback (Daft and Lengel, 1986). As a result, the development of relations with others can be impeded (Kiesler *et al.*, 1984) and take longer to reach the same level as co-located relations (Walther, 2002). While telephone conversations are not as rich as those that take place face-to-face (i.e., one cannot see facial expressions or body language), in comparison to text-based media, the telephone is superior as it does allow the transmission of verbal cues to indicate one's understanding and reactions, feedback is immediate and conversations tend to be lengthier (Axtell *et al.*, 2004). Moreover, the phone can help to 'transport' mobile workers into their familiar network of relationships in a form of 'absent presence' (Fortunati, 2002). Thus, given the relatively greater array of social cues in the telephone than text-based media, the formation and maintenance of relationships is likely to be easier. Therefore, we might expect that those mobile engineers who have greater reliance on text-based media might feel more isolated than those who are able to make more use of the telephone.

This effect is likely to be even more severe if those who have to rely on text-based media also have less face-to-face contact with colleagues because they lack a company base. Certainly, previous literature has suggested that socialization of teleworkers is easier when there is greater intra-organizational contact and regular use of real-time communications media like the telephone (Lamond *et al.*, 2003). This effect might also extend to feelings of isolation. In the following sections, we therefore examine this possibility among a sample of mobile engineers.

Sample and method

We interviewed 23 mobile engineers and their managers from four companies involved in photocopier or home appliance servicing and repair. From each company four or five engineers were interviewed along with one manager (although at one of the companies without a base two managers were interviewed). All interviewees worked in the North/Midlands area of England. Two companies had local bases from which the engineers collected parts, had their jobs allocated, their service meetings and from which they finished or ended their day at least twice a week. Jobs were collected from call controllers either in person when at the base or by telephone. The engineers also telephoned the call controllers when they had finished each job to let them know the job was complete. The other two companies (which were also the larger companies – both nationally distributed) no longer had local bases and the engineers started and ended every day from home. Parts were collected from post offices or other collection points and jobs were allocated electronically to engineers from a call centre or central office via a PDA or laptop. Engineers would tell call controllers that they had finished a job electronically via their PDA or laptop. Some phone conversations would occur with call controllers, for instance, if the engineer was running late on a job, but the technology reduced the number of phone calls required.

Semi-structured interviews were conducted which asked engineers about a range of issues relating to their job, including their contact and relationship with colleagues, use of technology, and the good and bad aspects of their work. All interviews were conducted in person and audio-recorded with the permission of the interviewees. They were fully transcribed and analysed for common themes using NVivo, which is a tool that aids the organization of data into themes.

Findings

The following section discusses the findings that emerged from the interviews. First, the similarities and differences in feelings of isolation between those with a base and those without will be illustrated with relevant quotes. Then, evidence will be presented as to why fewer differences emerged than might have been expected.

Similarities

Both those mobile engineers with and without a base could feel isolated during the course of their work as illustrated below:

> I do miss that quite a lot – there is no socialization most of the time, I am on my own listening to the radio – I do miss the chit-chat.
>
> (base)

> You are on your own. We have got people who we can speak to, but at the end of the day, the decision is yours.
>
> (no base)

Thus, there did appear to be a common element of 'being alone' when out on the road or dealing with customers for both sets of engineers.

Differences

Despite these similarities, three main differences occurred between those with a base and those without, in relation to opportunities for informal work interactions, social contact and building relations, and the integration of newcomers. These will be outlined in turn below.

1 Informal work interactions

Engineers with a base had more opportunities for impromptu chats and broader conversations about work as a result of being face-to-face. For example:

> the engineers all come in once a day and they all get together. I think you become a bit insular if you are out all the time and it is good for them all to get in and talk about the problems.... They might sit in here for half an hour in the morning. OK you might be having a laugh and a joke, but the main part of it is the jobs you did yesterday and the problems you came across, but it is done in a very loose social way.
>
> (base)

> Sometimes you have a conversation over the phone, but ... if you are face-to-face you go into other things ... and go into more detail ... about machines and so on ... [with a] phone conversation you tend to keep it to a minimum.
>
> (base)

Those without a base, however, had fewer opportunities for this:

> I think one thing that we have lost is the word of mouth to engineer, the group gathering in the morning.... The main thing they've lost is that contact with other engineers 'This job, that job' have a laugh, have a cry. That bit of it has gone. It is a very lonely, can be a very lonely job.
>
> (no base)

> I think the only difference between working in an office and remote is that ability to say 'Right everybody, quick meeting. I want to tell you about this' ... in the office it's a quick meeting [but] you're not able to get us together for two or three weeks.
>
> (no base)

As a result, conversations and meetings tended to be less frequent and more formal (or formally organized) among those without a base.

2 Social contact and relations

As well as discussions about work, those engineers with a local base were able to have social contact with colleagues when they met them in the office. For example:

> because we are in and out all the time ... we will come in for parts and if there are any engineers in you will sit and have a chat.
>
> (base)

> You can maintain [contact by phone], [but] I mean, one of the main reasons you do go to the office is so you can have that personal contact with them.
>
> (base)

In comparison, those without a base had relatively less social contact with others.

The use of PDAs and laptops to allocate jobs at these companies had particularly impacted on the relationship with the call controllers, as it had reduced the need for telephone contact with them:

> we used to have to call in [to the call controllers] maybe three or four times a day to pick up the [jobs] but now you might speak to them three or four times a week. The place could shut down and you would never know. So that is it as far as communication goes!
>
> (no base)

> [The call controllers] are just a voice on the end of the phone most of the time to be honest ... I've just got a new [call controller] but I don't know what she looks like and it could be several months before I'm down there to actually see who she is.
>
> (no base)

Relations with other engineers could also be less social:

> So technology is pushing us more away ... it is more impersonal and you are getting more and more alienated from your colleagues ... and they will talk about ... things as a team, and you think, 'What team?' I am not part of a team really ... I might talk to another engineer ... and I had been talking to him for four months and I did not know what he looked like.
>
> (no base)

> It can be a lonely job ... certainly working with work colleagues and getting to know people better – I mean, you can't arrange things for like a round of golf when you never see them.
>
> (no base)

Thus, the lack of base and use of technology did have some impact on social relations.

3 Integration of new engineers

It could be even more difficult for newcomers to get to know others when there was no central base. Apprentices at all companies had the benefit of a 'buddy system' where they accompanied another engineer for several weeks before going out on their own. However, this was not the case for fully trained engineers who transferred to another geographical area or from another company. The following quotes from engineers without a base illustrate the problems that newcomers could have developing social relations with colleagues:

> Well, we did get a guy who transferred from another area to [ours] but it's very occasional when you meet him really to be honest. We got introduced at a meeting. That was it really, and then you see him as and when.
>
> (no base)

> All that interaction side of it is lost … I think from their point of view, that's the ingredient that's missing. I think that probably applies to the new starters and apprentices and the like. I think some of the older experienced ones have learned to cope with it.
>
> (no base)

Thus, as a result of this lack of face-to-face social contact, new employees could only get to know their colleagues slowly over time. The lack of strong relationships to begin with could further impede the use of the phone as an alternative for relationship-building because engineers were reluctant to phone people whom they did not know well unless there was a specific reason:

> [I keep in touch] using my mobile, generally to ask for advice. I probably wouldn't just phone them for a social chat, probably because I'm quite new to the patch and I don't know most of them that well.
>
> (no base)

Lack of differences

Despite the illustrations above, there were not as many differences in relation to feelings of isolation as one might expect between those with a base and those without. In fact in general, problems of isolation were mentioned relatively infrequently by both sets of engineers, and tended to be most common for newcomers. Our interviews suggest that there are several reasons for the lack of differences found. First, those without a base tried to *replicate* the sort of contact possible through a company base by finding alternative locations for meetings, and using the mobile phone for support. Second, there were the *compensating factors* common to all the mobile engineers, i.e. the personal contact with customers and the freedom, autonomy and variety offered by being mobile. Finally, there were also '*engineer*' *factors* that played a role such as their personality and the long-term experience and tenure of many of the engineers. These issues will be illustrated in turn below.

Replication via the use of alternative location

One reason why those without a base did not feel much more isolated than those with a base was that other places were adopted for meetings, including impromptu face-to-face meetings. For instance, meetings took place around the parts collection points:

> an engineer can work on his own for weeks and weeks on end, and so we tend to … get little pockets of gatherings and usually it gathers round where the parts get dropped off … and several engineers attend probably first thing in the morning so they tend to see each other, so they have an informal chat which is very good.
>
> (no base)

> We all go to the post office to pick up our parts … and there's a canteen there and there's often four or five of us in the morning, so we'll go up and have a cup of tea and a chat … we do still meet … a lot of the lads do, you know, they'll go in half an hour before the shift starts.
>
> (no base)

Furthermore, engineers without a base would also have regular service meetings at convenient locations (such as hotels, village halls), mostly every four to six weeks, depending on need and workload:

> we have team meetings, you tend to get those religiously once a month…. We'll find a location somewhere in the district. It tends to be sports clubs, people use church halls, I use a village hall, you can find those sorts of places. You've got the guys then for four hours.
>
> (no base)

They would also try to simulate the impromptu work-related chats in those meetings:

> in the team talk … we always have the first bit as a 'free for all', everyone just has a go or whatever. But you know there is a serious element to it, in that someone might've come across a real strange problem and sussed it and they share it with everyone else, in case they come across it. So the first hour is just like a mad hour – you can say and do what you want.
>
> (no base)

Plus, there were occasional social events:

> we've grown … together … we've always maintained a social aspect of the job, so every now and again [the manager] will say … he'll put a message out and he'll say 'thanks lads for your efforts, it's been a busy week, well done, I think we deserve a night out' … and [he'll] buy a few drinks and that's reward enough for a couple of months.
>
> (no base)

Or in one of the companies the office came out to the engineers, in the form of the mobile training centre:

> we have a mobile training centre where we would do 'interactive training'. This is a big articulated lorry … because we don't have a base, the training comes to us.
>
> (no base)

Thus, the companies and the engineers themselves tried to replicate the advantages of the local base through other means.

Replication via mobile phone use

Despite the reliance on text-based media for job allocation, the mobile phone served as an alternative to face-to-face communications between engineers, and engineers and their managers. Although most conversations were work-related, there was also an element of social support involved. As illustrated in the quotes below, being able to contact others via mobile phone helped to reduce feelings of isolation.

> I never feel lonely … I can see that point, you know, that you think you're sat on your own … but I'm always speaking to engineers. Yes, it's not face-to-face, but … I wouldn't even have thought about it.
>
> (no base)

> Yeh, I think we just ring up for a chat … but it's usually a work issue to be fair. But yeh, we'll ring up just for a chat. After team meetings, often there'll be a lot of phone calls after team meetings, you know 'what did you think of that?' … But no, it's usually work, or kind of work-related let's say.
>
> (no base)

Managers in particular seemed to fulfil the role of social support for engineers and were often the people the engineers spoke to most. Indeed, part of the managers' role appeared to be to try to reduce the possible isolation of engineers by ensuring regular phone contact with them, and interviewees generally appreciated this. For example:

> [My current manager] tends to ring more than my [previous] manager, and he will just ring you up maybe not for any particular reason, just to see how you are. I would say [he] rings at least once a week or maybe twice a week . .. it is just nice to know that there is somebody there that is looking out for you.
>
> (no base)

Thus, the mobile phone could help to replicate at least some elements of social support that might occur at a local base.

Compensations in customer contact

Mobile engineers are not entirely on their own during the course of their work as they do meet customers and this was no different for those with or without base. Although the home appliance engineers had less opportunity to build relations, as they tended not to have the repeat visits, they saw relatively more customers a day and spent less time travelling on their own. Many engineers felt that contact with customers was one reason they did not feel isolated:

> I don't get lonely because I've built up, I've got my own customer base. I've built up quite a rapport with certain customers that we've got.
>
> (base)

> It is alright, because you are only on the road for 10 or 15 minutes and then you are back in someone's house. I can be a bit lonely I suppose, but because you are seeing people all day, working in their houses, it is not that bad as you still have people to talk to. You are not on your own throughout the day.
>
> (no base)

Compensations in freedom, autonomy and variety

A key issue that was mentioned frequently by engineers was the freedom, autonomy and variety in the job, which seemed to compensate for potential isolation. The two sides of the job (isolation and freedom/variety) were often spoken about together as two sides of the same coin. For instance:

> I hate being on my own and trying to cope with difficult jobs. Dealing with people on your own can be hard. But I do enjoy it. Every job is different and I am out meeting new people all the time.
>
> (no base)

Engineers especially commented on the flexibility they enjoyed by being mobile, particularly their ability to balance home and work life:

> I've also got in my home life … I've got a handicapped daughter and a handicapped wife.… So my presence of being local, field based and in my own destiny means that if necessary I can take time out between jobs to go and make sure everything's all right. For me that stops me moving to a different job.
>
> (no base)

> I think it is you are your own boss really you don't have to report to people – when you are out on the road you make decisions, you go into a customer's and you sort that machine out and it is basically left up to you – you have got flexibility if you want to call off and pick something up or call in the Bank or something. You are out on the road and you are not stuck in the same environment – every machine you go to it could be a difference fault.
>
> (base)

In addition, the fact that the engineers were not closely supervised and had the variety of working in different places rather than being stuck in the same office were seen as clear benefits of the work:

> It's not as if you're in an office where you've got someone who is forever looking over, 'oh you've not finished that yet?' sort of thing. You are, you know, working at your own pace – and although that is monitored we do have statistics, targets for the number of calls a day and everything – it's not the same as someone leaning over your shoulder all the time.
>
> (no base)

> I do like meeting different people all of the time and the variety that is given by the fact that I never work the same place for more than a day. I really like the variety. I don't think I could get up in a morning and go and sit at the same desk, with the same view out of the window.
>
> (no base)

Engineer factors: personality

There were many comments in the interviews about needing to have the right sort of personality to be a mobile worker. In particular being happy with one's own company and not requiring constant contact with others was considered important to combat the isolation. It was felt that engineers lacking such qualities would not last long in the job:

> at the end of the day, I suppose you have to have a temperament where you can work on your own. And in all honesty, I suppose I am that sort of person. I'm quite happy … you know how some people can't bear to be in a room on their own … with a book or without having the radio or TV on. I'm quite happy to be on my own.
>
> (no base)

> it is not suitable to everybody obviously. We have had engineers that have come and failed miserably because ... it is remote and they can't do with that, they prefer to be with people and contacts.
>
> (no base)

Engineer factors: experience, life-stage and tenure

Finally, one of the reasons that few differences were found between those with a local base and those without, was that most engineers we interviewed had been engineers for a long time and had a lot of experience. They did not need to ask for help very often and so did not feel such a great need for contact with others for support:

> it depends on the experience of the engineer, but certainly I only make a call [to my manager] maybe once a month, unless there is maybe other issues like [a recurring problem].
>
> (no base)

Moreover, because many of the engineers had been at the same company for several years, they had got to know the other engineers over that time period, and thus felt less need to build relations:

> we have known each other a long time so we don't need to see each other like every day or anything like that.
>
> (base)

Furthermore, some of those without a base had previously worked with the same colleagues before the bases had been disbanded, which had allowed them to develop stronger relations:

> round here ... most of the guys ... they were here when we had the [local] branch, so ... at that point I was seeing them most days. At least two or three times a week in the branch office.
>
> (no base)

Even without the benefit of a base, over time, relations could develop remotely. For instance, in relation to the call controllers, engineers were able to gradually build relations through phone calls and the occasional face-to-face contact:

> even though you never really see them face-to-face you might work with a controller – you have a designated controller at the call centre who deals with your calls for your area ... you have to speak to that one person, so you get a bond with that person.
>
> (no base)

Another issue about these, older, more experienced engineers, is that they were often at a life-stage where they did not feel the need to socialize with work colleagues:

> I think that as I have got older it is better because I have got a family life, but ten years ago, I missed the social life … because this is work basically.
>
> (no base)

> Maybe young people that are not in relationships that want a social life from their work might not enjoy this sort of work … when you work in an office you can go to the pub straight after with your colleagues and whatever, but in this sort of job you don't have that.
>
> (no base)

Thus, the age, life-stage and experience of the engineers appeared to reduce the problems of isolation, in part because relations had already been developed over time, and partly because less contact was required for technical and social support by these engineers. Concerns about isolation were more likely to be mentioned by engineers who were new to the job or geographical area.

Discussion and conclusion

In this study, although some feelings of isolation were felt by mobile engineers, it was mostly the younger, less experienced engineers or those who were new to the geographical area who felt it most keenly. For those without a base and who relied more heavily on text-based media, the problems of isolation could be more pronounced, but again, the impact was strongest on newcomers (particularly with regards to building social relations and integration). This might be considered surprising given that previous literature found isolation to be an issue (i.e. in Laurier and Philo's, 2003 ethnographic study). However, their 'lone employee' was relatively new to her job and so this may be one reason explaining why her loneliness seemed so severe.

The majority of engineers in the current study were protected from isolation through *replication* (by meeting at alternative locations and using the mobile phone for support), *compensation* (contact with customers and the relative freedom/variety in the job) and *characteristics of the engineers* themselves (personality and experience/life-stage). For instance, the mobile phone, with its greater array of social cues relative to text-based media, can replicate less formal social interaction between those who know each other well, and may also help to compensate for being 'absent' from familiar surroundings through enabling a 'virtual presence' with one's colleagues (Fortunati, 2002). Even those who relied more on text-based media were able to use the mobile phone and could gain some 'social' compensation from it.

Replication also occurs regarding the location of work, with alternative spaces being adapted to recreate the local office. This is reminiscent of Brown and O'Hara's (2003) assertion that mobile work 'makes place', such that it transforms traditionally non-work locations (such as village halls, cafes etc.) into workplaces. Other authors have also commented on how mobile workers appropriate different (often public) spaces for work purposes (e.g., Felstead *et al.*, 2005; Forlano, Holm and Kendall, in this volume). Being creative with space becomes a necessity in mobile work, particularly for those without a traditional office base who might not otherwise be able to interact with colleagues face-to-face. Moreover, replicating the informal interaction that can occur in chance meetings at the office also becomes a key goal. This might be through frequenting places that other colleagues are likely to visit or by allowing time for free conversation during formal meetings. Such practices and behaviours appear to go some way towards protecting mobile workers from isolation. Isolation from colleagues in this sample was compensated for via interaction with customers. The frequent contact with customers (in several cases the customers were well known to the engineers) prevented them from feeling as 'alone' as they might otherwise have done. This effect might be job-specific. For instance, in Laurier and Philo's (2003) study, the 'lonely' mobile worker was a salesperson, a role likely to involve quite different interactions and relations with customers than might occur in service jobs where customers require the help of the employee and may be more pleased to see them. Thus, contact with customers may be less of a compensating factor against isolation in other occupations such as sales.

The other compensating factor was the relative freedom and variety offered by being mobile. Certainly, in comparison to other forms of telework, mobile teleworkers enjoy greater variety of location (Daniels, 2000). This variety can also increase the mobile worker's freedom to a certain extent (e.g. their work-based route may pass a bank or shop where they can quickly conduct some personal business). So, in this respect, mobile workers may have advantages over other types of teleworker.

Finally, personal characteristics also offer protection from isolation in mobile work. The idea that it takes a certain sort of person to be a teleworker is not new (Lamond *et al.*, 2003) and this seems to extend to mobile workers. Being happy and capable of being alone certainly seems to help, as does being at a certain life-stage and level of experience where social support from work is less important.

Thus, in comparison to home-based teleworkers, these mobile teleworking engineers appeared relatively isolation-free and did not suffer the same problems of work–life balance. Indeed, work–life balance appeared to be somewhat facilitated by the freedom associated with being mobile. Nevertheless, despite these seemingly positive findings, the interviews do indicate that caution is required with new hires when mobile, particularly without the benefit of a local base (and when compounded by reliance on text-based media). Socialization is especially problematic when the level and synchronicity of communication is low (Lamond *et al.*, 2003). In these circumstances extra effort is required to try to integrate engineers and help them to develop relations with their colleagues as otherwise attempting replication of social contact via technology may be less successful or take much longer. This is because, if engineers do not know their colleagues well it can be much harder to make use of the mobile phone to gain social support from them. The remote manager's role may be particularly important in this regard, not only for offering that support themselves (which is often seen as part of their role anyway), but as boundary-spanners helping colleagues to develop their networking skills and relations with each other (Duarte and Snyder, 2001). Thus, encouraging interaction between colleagues and providing opportunities for them to interact is also a key part of their role.

So, in conclusion, being mobile, lacking a company base and relying on communications technology need not lead to a 'lonely life' as long as there are means to replicate and compensate for the lack of face-to-face contact and employees are at a stage where they require less social contact with colleagues. Each type of mobile job may have different levels of these features which may need to be balanced. Some factors (such as customer contact) may be less effective at compensating for isolation in some mobile jobs, although other compensating factors may be available. Studies of different types of mobile work are necessary to gain a better understanding of this issue. Moreover, given that the main difficulties seem to occur for newcomers, more research on relationship development in new mobile workers would be warranted.

References

Axtell, C. M., Fleck, S. J. and Turner, N. (2004) 'Virtual Teams: Collaborating across Distance', in C. L. Cooper and I. T. Robertson (eds) *International Review of Industrial and Organizational Psychology*, vol. 19. Chichester: Wiley, pp. 205–247.

Belanger, F. (1999) 'Communication Patterns in Distributed Work Groups: a Network Analysis', *IEEE Transactions on Professional Communication*, 42: 261–275.

Brown, B. and O'Hara, K. (2003) 'Place as a Practical Concern of Mobile Workers', *Environment and Planning A*, 35: 1565–1587.

Churchill, E. and Munro, A. (2001) 'Work/Place: Mobile Technologies and Arenas of Activity', *SIG Group Bulletin*, 22/3: 3–9.

Daft, R. L. and Lengel, R. H. (1986) 'Organizational Information Requirements, Media Richness and Structural Design', *Management Science*, 32: 554–571.

Daniels, K. (2000) 'Job Features and Well-being', in K. Daniels, D. Lamond and P. Standen (eds) *Managing Telework: Perspectives from Human Resource Management and Work Psychology*. London: Thomson Learning.

Duarte, D. L. and Snyder, N. T. (2001) *Mastering Virtual Teams.* San Francisco, CA: Jossey-Bass.

Felstead, A., Jewson, N. and Walters, S. (2005) *Changing Places of Work.* Basingstoke: Palgrave Macmillan.

Fortunati, L. (2002) 'The Mobile Phone: Towards New Categories and Social Relations', *Information, Communication & Society*, 5: 513–528.

Haws, U. (1984) 'The New Homeworkers: New Technology and the Changing Location of White Collar Work', *Low Pay Pamphlet No. 28* (London: Low Pay Unit).

Hislop, D. and Axtell, C. (2007) 'The Neglect of Mobility in Contemporary Studies of Telework', *New Technology Work and Employment*, 22: 34–51.

Kiesler, S., Siegel, J. and McGuire, T. W. (1984) 'Social Psychological Aspects of Computer-mediated Communication', *American Psychologist*, 39: 1123–1134.

Lamond, D., Daniels, K. and Standen, P. (2003) 'Teleworking and Virtual Organisations: The Human Impact', in D. Holman, T. D. Wall, C. W. Clegg, P. Sparrow and A. Howard (eds) *The New Workplace: A Guide to the Human Impact of Modern Working Practices*. Chichester: Wiley.

Laurier, E. and Philo, C. (2003) 'The Region in the Boot: Mobilising Lone Subjects and Multiple Objects', *Environment and Planning D: Society and Space*, 21: 85–106.

Mann, S. and Holdsworth, L. (2003) 'The Psychological Impact of Teleworking: Stress, Emotions and Health', *New Technology Work and Employ*ment, 18: 196–211.

Sproull, L. and Kiesler, S. (1986) 'Reducing Social Context Cues – Electronic Mail in Organizational Communication', *Management Science*, 32: 1492–1512.

Walther, J. B. (2002) 'Time Effects in Computer-mediated Groups; Past, Present and Furtue', in P. Hinds and S. Kiesler (eds) *Distributed Work*. Cambridge, MA: MIT Press, pp. 235–259.

9 Re-space-ing place

Towards mobile support for near diagnostics

Mikael Wiberg

Introduction

During the last ten years we have been able to witness a fast development towards IT-supported physical environments. This development has included both the development of mobile technologies that we can carry with us in various settings, as well as the development of sophisticated sensor technologies that we can use to digitalize our physical surrounding.

This research focuses on the industrial setting as one example of a physical surrounding that is undergoing a development towards this digitalization. With this current development in focus, this study is interested in understanding how we can supply the individuals working in these settings with appropriate IT support. Given this interest, the research approach is user-centred, and design-oriented in that it takes a point of departure in empirical studies of the people working in an industrial setting for the purpose of identifying new and novel ways to support these people with novel IT. As such, this study focuses on the operator's ongoing work of diagnosing and serving the machines and equipments at the research site, i.e. one of Boliden's factories in Sweden.

The rest of the chapter is structured as follows: It first presents an overview of current research of digitalization of environments with a specific focus on industrial settings. The chapter then explores the motivation behind this research, going on to describe the research method applied and the research site of Boliden. The main findings from this empirical study are extracted and presented, followed by a discussion of some possible design implications indicated by the results. Finally, the overall design of two prototypes, intended to support near diagnostics for mobile operators, are illustrated followed by a discussion which takes one step back to reflect upon this new IT support in terms of near diagnostics, and in terms of a 're-space-ing of place' from the *in situ* sense-making perspective of the mobile operators.

Background

Today, we can easily find many examples of the current digitalization of our physical surroundings. Cars, for instance, used to be solely mechanic constructions. These vehicles can now contain around 50 microprocessors, several computer networks and a lot of digital sensors to support the driver and provide good comfort and safety.

Another area is the industrial setting. In the last decade a lot of effort have been put into the digitalization of the traditional physical and mechanical industrial setting. Here, the guiding and unifying concept has so far been 'remote diagnostics', focused mainly on the possibilities of placing sensors around the industrial plants to enable the abstraction of real-time data from the physical machines in the factories, and the potential to transfer these data *away* from the physical location where they are produced to some other place (e.g. a remote-control room) for further computation, analysis and storage. Examples of research examining this development include Biehl *et al.* (2004), who adopt a broader perspective on remote diagnostics in their research into remote repair, diagnostics and maintenance; and Jonsson and Holmström (2005), who discuss remote diagnostics in terms of ubiquitous computing to capture this ongoing development towards the digitalization of physical surroundings.

To realize this development, we have during the last decade developed a wide range of technical facilities enabling us to digitalize physical places and machines. These enablers include microprocessors, sensors (temperature, heat, vibrations, pressure, light, sound, etc.), computer networks, good batteries, and RFID technology, including both sensitive readers and RFID tags especially developed for rough environments.

Looking at remote diagnostics in more general terms, it is quite easy to see how it boils down to the *abstraction of information* from machines, processes and activities at a specific location, together with the *transportation of information away* from the machines producing this information, and *away* from the mobile operators working with these machines.

One might even read into this development a Tayloristic view of the factory, i.e. a development towards a dream about the self-diagnosing, self-operated factory with no mobile operators (i.e. humans) present in the plant. Even if this dream might still include people present in the factory, the main idea is still to relocate the intellectual problem-solving, analysis and diagnosis from the operators working with the machines to expert analyzers at a remote location.

Motivation

While a lot of attention has been directed towards the design of technologies to support *remote diagnostics* in industrial settings, this research explores a complementary path, i.e. mobile support for *near diagnostics*. Here, near diagnostics can be defined as '*the individual,* in situ *sense-making and actions taken to adjust, improve or correct industrial machinery as to make the whole production process more smooth, effective and safe*'. This focus is primarily motivated by the problems that have been identified with a single development towards remote diagnostics (including not least the relocation of intellectual problem-solving as highlighted above), but this focus is also motivated by an intellectual desire to explore what a contrary development to remote diagnostics could contain and if it could be a valuable complement to the mainstream research in this area. Near diagnostics was also a guiding concept that meriting further explanation and validation on this empirical study, i.e. the aim was to see if any activities taking place in the industrial setting could be labelled as 'near diagnostics' and if, based on these observations, any ground for a further exploration of IT support for near diagnostics could be found.

This chapter's overall interest concerns understanding how IT can be designed to support *in situ* sense-making and actions taken from the perspective of near diagnostics. This interest, in relation to the current research into remote diagnostics, represents a new focus, at the same time as it is grounded in, and inspired by, some important previous research including Orr's (1996) work on doing ethnographic research into the service and maintenance of machines; Suchman (1987) on the fundamental theory about situated actions and *in situ* sense-making, which is very valuable to the work in this area. This research is also inspired, and motivated by Hutchins (1995) on the theory of distributed cognition, in which he specifically highlights and acknowledges the important interplay between the individual and other individuals, the physical surrounding, and artefacts in the physical surrounding for effective problem-solving and *in situ* sense-making. Finally, this research also represents a continuation of my previous research into mobile support for mobile operators, including mobile IT support for service technicians (Wiberg, 2001).

The next section describes the method applied in this project, before moving on to the research site at Boliden.

Research method

In order to reach the goal of identifying new ways to support mobile operators' work with novel mobile IT, and to reach our second goal of exploring and validating the concept of near diagnostics, the research approach was designed as follows.

Ethnographic techniques (Hammersley and Atkinson, 1993) were employed to collect data to identify, capture and gain an understanding of the mobile operators work, including their daily routines, patterns, habits, and everyday work with the machines in the plant. On-site observation of the operators, during their rounds through the plant, was conducted. In general terms, ethnography is about

> participating in people's daily lives for an extended period of time, watching what happens, listening to what is said, asking questions – in fact, collecting whatever data are available to throw light on the issues that are the focus of the research.
>
> (Hammersley and Atkinson, 1993)

The objective of ethnography is, further, to 'present a portrayal of life as seen and understood by those who live and work within the domain concern' (Hughes *et al.*, 1994). Thus, presentation of the results from this phase of the study will mainly consist of descriptions of how work is carried out in relation to maintenance and the concept of near diagnostics.

Based on this study, an interaction model was developed to set the scope for the design phase of the research project. Equally important, this development of a conceptual model of the mobile operators' work with the machines, and with their IT upport, enabled researchers to identify, exemplify and visualize an important design and conceptual gap from the perspective of near diagnostics and *in situ* sense-making. To move on to the goal of identifying new ways to support mobile operators' work with novel mobile IT the next phase of concept development tried to identify interesting aspects of their work to target in the design, and to develop interesting concepts and metaphors that could be communicated back to the mobile operators. This communication part was important in order to validate the study's conceptual ideas in the realm of their work setting and could be seen as a first evaluation of our design proposals. Finally, two conceptual prototype systems based on the concepts were developed in order to validate the possibilities of providing the mobile operators with IT support for near diagnostics.

The next section begins with a background description of the research site at Boliden before delving deeper into the empirical study conducted at the industrial site with the mobile operators.

Research site: Boliden

Boliden is one of the leading mining and smelting companies in Europe with operations in Sweden, Finland, Norway, Ireland and the Netherlands. Boliden's main products are copper, zinc, lead and gold and silver. The number of employees is approximately 4,500 and the turnover amounts to approximately €3.8 billion annually.

Boliden is the third largest copper producer and the second largest zinc producer in Europe. By being one of the leading companies in the industry in terms of responsibility, reliability and customer satisfaction, Boliden intends to attain its overall goal of becoming Europe's foremost copper and zinc producer.

The company's operations focus on the initial stages of the processing chain, in other words, exploration, mining and milling, smelting, refining and recycling (see Figure 9.1). Metal recycling is a field in which Boliden is a global leader and it also represents a growing sphere within its operations. The company conducts operations in the following business areas: mines, smelters and market.

The next section reports results from the ethnographic field study at the company's plant in Boliden, Sweden.

Results

The plant at Boliden is a complex construction consisting of a huge number of machines, pumps and other moving parts (see Figure 9.2 for an overview of the

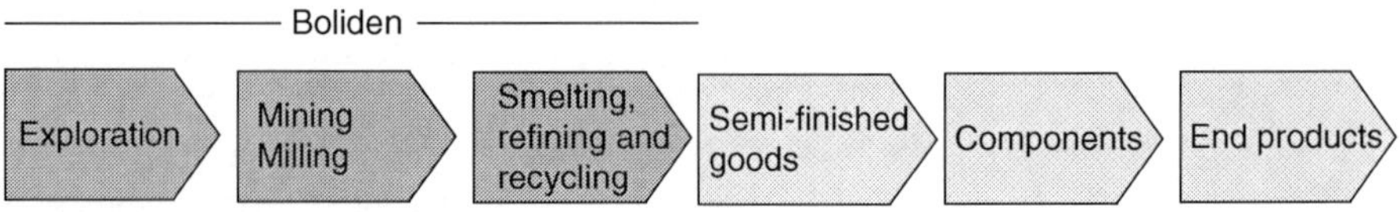

Figure 9.1 Boliden's operations focus of the processing chain.

Figure 9.2 Overview of the Boliden plant.

plant). In order to get a daily overview of this complex construction, the mobile operators rely on both their own senses as well as sensory technologies, including sensors to measure pressure, flows, temperature, etc. at a huge number of physical locations. These sensors are attached to the machines and connected to the computers in the control rooms. This arrangement allows the operators to get hands-on information about the machines' conditions by being out in the plant and listening for strange noises, and touching the machines to feel vibrations, etc. They can also observe flows and other indicators of how the process is going. At the same time, they can enter the control room to check the data generated from the machines and sensors, and to look at more long-term reports and curves that are constantly generated by the process control system.

The whole plant is organized into several gates, which makes it easy for the mobile operators to move around quite freely to check, diagnose and serve the machines. While the physical layout of the plant enables them to move around freely, they do at the same time also have the responsibility of carefully checking the whole process and its components several times a day. To facilitate this, their work is organized into daily routines and rounds through the plant. During these rounds, the mobile operators follow a specific path through the factory, looking and listening for problems and taking notes about anything needing attention or further analysis.

Since Boliden is a process industry, it is crucial that the processing (and thus the production of zinc and copper) is (almost) never stopped due to the high costs of an unplanned stop. Instead, planned production stops when many minor problems can be taken care of simultaneously during a short period of process downtime, represent the ideal aim. From the view of the mobile operators, this means that it is seldom an option to immediately take a machine, a pump or some other component out of the process, fix it or replace it, and start up the whole process again. Instead, the operators need to take notes and prioritize necessary jobs for the next planned stop.

The control room (see Figure 9.3) provides operators with a good working environment for controlling the plant via the computers, offering a good overview of the running processes via the sensor data that is constantly fed into the system. At the computer monitors, the operators can follow any change in the plant, get warning alerts, and look at the reports generated by the system. They also have a live video-feed from several locations within the plant so that they can double-check the data with a live video from crucial areas (see the

Figure 9.3 The control room at Boliden.

small black monitor in Figure 9.3). Here in the control room the operators analyze the process and double-check the things noticed during their rounds through the plant.

However, this double-checking of the data in the control room is seldom enough to get a full understanding of changes in the plant. Instead, the mobile operators are typically forced to return to the machines generating the alarms or malfunctioning data in order to fully make sense of the situation and decide on action. This is what this chapter labels as 'near diagnostics', i.e. *the need for a close-up, on-site, and hands-on analysis of a machine's condition*. Near diagnostics is thus about going the other way compared to remote diagnostics, i.e. instead of looking at a machine and having digital sensors and computer networks to transfer this image of the process away from the machines to a remote location for further analysis, near diagnostics is about moving out from the control room, and back to the machines that generate this data for on-site analysis and making sense of a malfunctioning process, scaffolded by the data image of the malfunctioning component. In a way, this is similar to the need for an embodied understanding of, and embodied interaction with, a malfunctioning process as pinpointed by Dourish (2001) in his book *Where the Action Is*.

Figure 9.4 shows three different examples of this near diagnostics activity taken from the empirical study at Boliden. Figure 9.4 (top) shows a transportation band with cooper. The system indicated to the operators in the control room that the band was moving too fast in relation to the rest of the process. However, when on site, the operator could see that, although the band was moving a little bit faster than normal, it was not causing any problems at the next station in the process. In the next figure (Figure 9.4, centre) another alarm indicated that the pipe in the rear had too fast a flow for the copper container. Here, the mobile operator was aware of this problem when arriving on site and he could easily adjust the flow to a more appropriate level. He then needed to return to the control room to get the exact data about the new flow through the pipe. The third example (see Figure 9.4, bottom) concerned a vibrating pump noticed by the mobile operator during a round through the plant. Here, there were no indications about this in the control room (since it is almost impossible to put sensors on every moving part of the plant). However, the operator returned to the control room to check that the process was still okay, making a note about this pump and adding it to the service schedule for the next planned production stop.

These three examples illustrate, in different ways, the need for near diagnostics activities, with the mobile operators required to move back and forth between the machines and the control room. In some cases, they need to return to the control room to double-check the process in relation to the sensor data, and in other cases they are pre-informed about a situation via the sensor data, which makes them well prepared for an on-site, close-up analysis of any component of the factory. As such, the mobile operators' overall behaviour can be described as a constant moving in and out of the control room (compare the classic article 'Moving out of the Control Room' by Hughes *et al.*, 1994).

To relieve this problematic situation for the mobile operators of having to

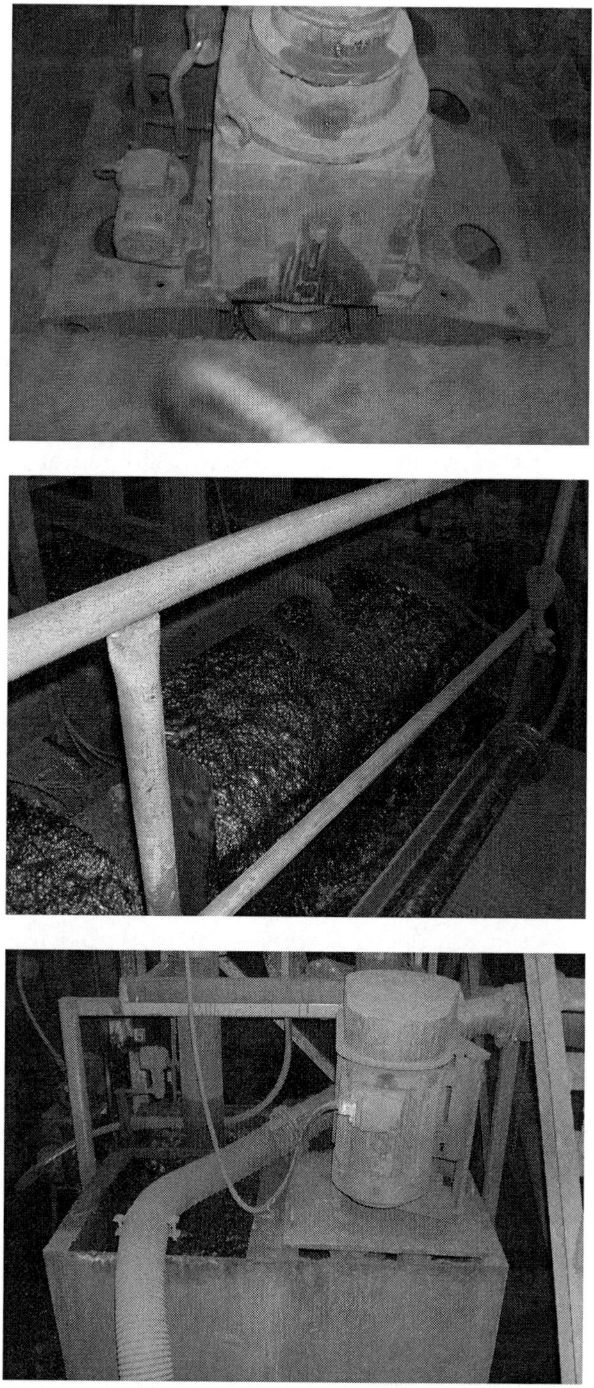

Figure 9.4 Three different examples of near diagnostics at Boliden.

transport themselves back and forth between the control room and the machines, two solutions have been adopted. First, they have equipped themselves with traditional communication radios so that they can talk with each other when one operator is in the control room and another in the plant. However, this solution can in many situations be cumbersome since it is hard to describe many of the complex installations made in the plant over the radio, and, from the other way around, it can be hard for the operator in the control room to localize a particular sensor in order to help and guide a colleague operator on the floor. The second solution for tighter, more short-range access to the data consisted of a small control room placed in the middle of the plant (see Figure 9.5). Due to its central position, the operators typically have a shorter walk to this simple control room than to head all the way back to the main control room next to the main entrance of the plant. Here, the operators can pop in and check the computer monitors, and then go back to the machines again. Figure 9.5 illustrates one such situation, with a mobile operator checking some data in the process control system.

While this small control room lessens the distance between the machines producing the sensor data and the access to this data, the distance is still an issue since the individual operator cannot do on-site adjustments to the machines and follow how it affects and changes the values in the system. Further on, this geographic separation of the machines and the data also creates a temporal gap between observing a running process, and then, at a later stage, checking the data generated by that process in the control room.

So, while the everyday actions of near diagnostics is an empirical fact, we have also seen that the mobile operators studied lack technical *in situ* support for guiding their mobile calibration work and their on-site sense-making. Instead, they typically have to return to the central control room whenever they need to check any sensor data from the machines that they are serving, a phenomenon we have come to label as micro-routes.

The results from this empirical study thus confirm the hypothesis about the existence of near diagnostics activities and the need for a better integration

Figure 9.5 The small control room at Boliden.

between the machines, the mobile operators and the data produced by the sensors in the plant. Research revealed that they had tried to work around this gap by using communication radios and the small control room, but that the gap still existed despite these efforts. Further on, the moving in and out of the control room behaviour we think about as a two-step cycle in which the moving out on a round is about collecting personal sensory data about the processes and machines, whereas the moving in round is about analysing the data collected and trying to make better sense of their *in situ* analysis in relation to the overall production process.

Based on these observations, the next section outlines an initial model of the identified gap between the machines, the operators and their IT support in order to gain an informed overview of how appropriate IT support for near diagnostics could help these mobile operators.

Constructing an interaction model for near diagnostics

Based on the empirical study as reported above, and by following a human-centred approach, taking a specific point of departure in the individual mobile operator and how to support them with novel IT, a threefold gap can be identified in relation to the mobile operators, the machines they are serving, and their supporting IT infrastructure. In my view, this threefold gap includes 1) *a geographic gap* (between the machines they are serving out in the plant, and the physical location in which the computers are located, i.e. the control rooms, 2) *a logical gap* (related to the separation of information generated by the machines from the machines that are actually generating this information, and finally 3) *an interaction gap* (in that the interaction with the machines are separated from the interaction with the computers, and vice versa). If structuring this into a model (see Figure 9.6), one can see a triangular relationship between the mobile operators (people), their IT support (IT) and the machines they are diagnosing and serving (machines).

This threefold gap is of course a delicate problem. On the other hand, it presents an interesting design space for IT support for near diagnostics if extending the model as follows. Given the identified gaps, this study envisions, first, an interaction model that takes a point of departure in the individual mobile operator (see 'people' in Figure 9.6); and second, that the mobile operator should be

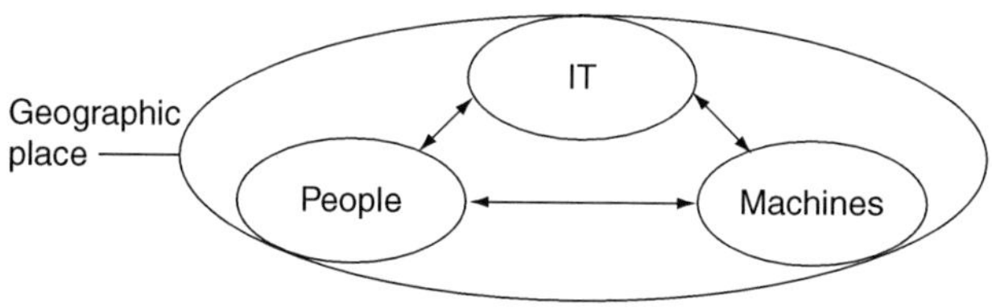

Figure 9.6 Interaction model of the triangular relationship between the mobile operators, their IT support, and the machines they are serving.

able to interact with the machines, as well as the computers at the same physical location (see the circle labelled 'geographic place' in Figure 9.6), i.e. in front of the machines out in the plant. Finally, it envisions a tight integration between the interaction with the machines, the interaction with the computers, and the link between the machines and the computers, so that actions taken to adjust the machines is reflected on the computers and vice versa.

To realize this vision, and to explore this design space, the research has taken this general interaction model of near diagnostics as a guiding framework for targeting different aspects of IT support for near diagnostics, and for identifying a couple of design implications or guidelines that should be considered in order to develop good IT support for the mobile operators that could bridge these identified gaps. These guidelines and implications for design are outlined below.

Based on this empirical study, and guided by the interaction model for near diagnostics as outlined above, I have come to the conclusion that IT support for near diagnostics should be 1) *mobile* – as to allow for access to the sensor data anywhere in the plant, 2) *lightweight* – in both physical terms as well as interaction wise, since the mobile operators already carry a lot of equipment and tools on their rounds and, if standing in front of a machine, their focal attention should be directed towards the machine to be analyzed rather than any complicated user interface. It should also be 3) an *integrated mobile–stationary solution*, so that the mobile operators can move seamlessly between the control room displays and the mobile device that can be carried out in the plant. Finally, the mobile operators are used to working with their bare hands when repairing and serving the machines, so we believe that a solution that builds upon 4) an *embodied interaction modality* could be appropriate to support them in their daily work.

Towards mobile support for near diagnostics

While remote diagnostics is mainly about 'anytime, anywhere' computing in its focus on abstracting and transporting information *away* from its geographical origin to some other location, the focus here on near diagnostics is something contrary to this idea. We think that, by augmenting the physical location with place-related information, we can in fact support a sense of a place while being there instead of forcing the mobile operator to move away from the situation at hand to access a computer in a remote control room. This project has therefore considered two prototypes as two different conceptual illustrations of mobile support systems that could enable mobile operators to put a specific focus on the physical location at hand.

MoveInfo – pick and drop from a large control room display

The first prototype is called MoveInfo, and the basic idea behind this prototype was to provide the mobile operators with a simple way of transferring real-time data from a big control room display onto a small, portable device that could be carried out in the plant.

Inspired by the 'pick-and-drop' interaction technique developed by Rekimoto

(1997), MoveInfo allows the mobile operator to select certain pieces of information from the big control room displays by using a stylus to click on the important information. The operator can then use the stylus to drop this information onto the portable device so that this information is available to him in the plant but still also available in real-time, thanks to a WiFi connection to the process control system in the plant (see Figure 9.7).

From a research perspective, the application domain of pick-and-drop can be expanded from being a tool to support office work and informal meetings to the industrial use setting.

Fallman *et al.* (2005), were responsible for a similar contribution, applying pick-and-drop to the industrial setting to enable a wall-size view of information located at mobile devices.

GeoInfo – in situ *place-associated data for process maintenance*

The second prototype is called GeoInfo. Here, the basic idea was to provide the mobile operators with a portable device able to function almost like a traditional magnetic compass, in that it is aware, not of direction, but of location, and can thus continuously supply place-based information about any machines currently close to the operator (including pump pressures, volumes, temperatures, vibrations, etc.) by sending the sensor data normally only sent to the central control room to the mobile device as well. This installation enables the operator to move around in the plant and observe the machines from a first-person perspective while having access to the sensor data at any location. It also allows the operator to choose to follow a certain flow, route or process through the plant to discover and make sense of chain-reaction problems caused by other machinery along the process. Figure 9.8 illustrates this installation.

From a research point of view, this expands the previous work on the

Figure 9.7 A mobile operator uses a stylus to pick information items from a big control room display and drop it onto a mobile device.

Figure 9.8 A mobile operator serving a machine while having access to the sensor data produced by the machine in front of him.

GeoNotes system by Persson *et al.* (2002) through the application of the compass metaphor, which suggests a fully automatic and continuous update of the portable device, depending on the current location. From the perspective of the user, this also means that the operator does not need one hand to work the portable device, nor to find the machine in front of him in any complex menu system. Instead, the device always shows the sensor data from the machines proximate to the mobile operator.

The next section elaborates on how these prototypes could change the work situation for the mobile operators from a sense-making perspective.

Discussion – A re-space-ing of place?

So, how might the two prototype systems presented here change the work situation for the mobile operators from a sense-making perspective? Well, first of all, we believe that the two systems illustrate two different ways of bridging the threefold gap as identified and outlined above, and in doing so, these systems enable the mobile operators to have their whole sense-making process (including both hands-on experience and data analysis) located in front of the machines.

On a practical level, this analysis of IT support for mobile operators contributes to previous work conducted by Orr (1996), in that it focuses specifically on local mobility as a fundamental aspect of service and maintenance work with machines. And, on a more theoretical level of analysis, the design-oriented efforts presented in this chapter contribute to the work done on situated actions by Suchman (1987), in relation to her reflections on 'material resources for situated actions', from the viewpoint that the tools being explored provide new means for better-informed *in situ* actions for the mobile operators working at the Boliden plant. This also leads back to the work on distributed cognition by

Hutchins (1995). Here we can think about the prototype systems in terms of mobile, and yet environment-related or environmentally aware resources for collective problem-solving processes, especially if putting the two system prototypes proposed here in the context of a collaborative situation including both the mobile operators as well as the control room personnel and their shared task of analysing, diagnosing and maintaining the machines.

This thinking has been inspired by the 'Re-place-ing space' paper by Harrison and Dourish (1996), in which they elaborate upon the notion of place vs space. According to them, '*space is the opportunity; place is the understood reality*' (p. 69); and '*physically, a place is a space which is invested with understandings*' (p. 69). In relation to our description of 'remote diagnostics' vs IT-support for 'near diagnostics', we interpret this as follows: In remote diagnostics, the data for analysis (and thus understanding) is transferred to another location, which in fact limits sense-making at the site, i.e. a deconstruction of a place into a space. Harrison and Dourish (1996) then continue arguing that: '*spaces are part of the <u>material</u> out of which places can be built*' (p. 73). In our view, our proposed systems constitute material that can be brought into a location, bridge the threefold gap, and added to a space so as to construct a place invested with understanding, i.e. a material support to scaffold *in situ* sense-making. In accordance with this thinking, we view our efforts as a 're-space-ing' of the mobile operators' places by bringing back the sensor data to the places that generate this information. From this perspective, 're-space-ing' of places becomes a design space to be explored and we conduct this exploration through the augmentation of physical locations using lightweight and mobile computational power and easy access to sensor data. To us, the re-space-ing of places presents us with a new design space and opportunity to elaborate with mobile IT as a design material.

Conclusion

This chapter has presented an empirical study of mobile operators in an industrial setting, with the goal of identifying new ways to support their work with mobile IT. The study confirms the hypothesis about the existence of near diagnostics activities and the need for a better integration between the machines, the mobile operators and the data produced by the sensors in the plant. It was evident that they had tried to work around this threefold gap via communication radios and creating a small control room, but that the gap still existed despite this support. To bridge this gap, this chapter outlines two prototype systems enabling mobile operators to focus on the physical space where they are working, constituting a way of 're-place-ing space' through supporting *in situ* sense-making.

Through this confirmation of near diagnostics as an important activity in the factory, we can also draw the more general conclusion that people are still needed in the factories (despite any Tayloristic visions of the fully automated plant).

This chapter also extends previous research on the use of pick-and-drop and GeoNotes to a new application domain, which also serves as a good validation of previous research.

Finally, this project has demonstrated an important lesson when it comes to understanding and supporting mobile work, i.e. that one needs to pay specific attention to the mobility patterns of people, in this case how the mobile operators at Boliden constantly move in and out of the control room, and their need for near diagnostics to carry out their daily work.

Acknowledgements

The research reported in this paper has been conducted as part of the 'Process Information for Near Diagnostics' project, which is part of the ten-year VINNOVA-funded project ProcessIT, www.processitinnovations.se/.

References

Biehl, M., Prater, E. and McIntyre, J. (2004) 'Remote Repair, Diagnostics, and Maintenance', *Communications of the ACM*, November, 47: 11.

Dourish, P. (2001). *Where the Action Is*. Cambridge, MA: MIT Press.

Fallman, D., Kruzeniski, M. and Andersson, M. (2005) *Designing for a Collaborative Industrial Environment: The Case of the ABB Powerwall*. Proceedings of DUX 2005 Conference on Designing for User Experience, San Francisco, CA: ACM Press.

Hammersley, M. and Atkinson, P. (1993) *Ethnography: Principles and Practice*. London: Routledge.

Harrison, S. and Dourish, P. (1996) *Re-place-ing Space: The Roles of Place and Space in Collaborative Systems*. Proceedings of the 1996 ACM Conference on Computer Supported Cooperative Work.

Hughes, J., King, V., Rodden, T. and Andersen, H (1994) *Moving out of the Control Room*. Proceedings of the 1994 ACM Conference on Computer Supported Cooperative Work, pp. 429–439.

Hutchins, E. (1995) *Cognition in the Wild*. Cambridge, MA and London: MIT Press.

Jonsson, K. and Holmström, J. (2005) *Ubiquitous Computing and the Double Immutability of Remote Diagnostics Technology: An Exploration into Six Cases of Remote Diagnostics Technology Use*. Proceedings of the IFIP 8.2 Conference.

Mason, R. O. (1989) 'MIS Experiments: A Pragmatic Perspective', in I. Benbasat (ed.) *The Information Systems Research Challenge: Experimental Research Methods*, vol. 2. Boston, MA: Harvard Business School Press, pp. 3–20.

Orr, J. (1996) *Talking about Machines: An Ethnography of a Modern Job*. Ithaca, NY: ILR Press, Cornell University Press.

Persson, P., Espinoza, F., Fagerberg, P., Sandin, A. and Cöster, R. (2002) 'GeoNotes: A Location-based Information System for Public Spaces', in K. Höök, D. Benyon and A. Munro (eds) *Readings in Social Navigation of Information Space*. New York: Springer.

Rekimoto, J. (1997) 'Pick-and-Drop: A Direct Manipulation Technique for Multiple Computer Environments', proceedings of UIST 1997, pp. 31–39.

Suchman, L. (1987) *Plans and Situated Actions: The Problem of Human–Machine Communication*. Cambridge: Cambridge University Press.

Wiberg, M. (2001) 'In between Mobile Meetings: Exploring Seamless Ongoing Interaction Support for Mobile CSCW', PhD thesis, Sweden: Department of Informatics, Umeå University, Umeå.

10 420 years of mobility

ICT-enabled mobile interdependencies in London hackney cab work

Silvia Elaluf-Calderwood and Carsten Sørensen

Introduction

For 420 years it has been possible to hire someone to drive you around London by simply going to a cab rank or hailing one on the streets. The mobile workers conducting this activity have since 1622 been recognized as licensed hackney carriage drivers. Since 1851, drivers have been required to study for a licence called 'The Knowledge', for which they intensively study over 300 routes covering 25,000 roads and destinations in central London. The solitary mobile working of London cabbies has traditionally also been highly individual work where drivers themselves decide where and when they work. However, the advent of mobile phones, Global Positioning Systems (GPS) and computer-cab dispatch technology, along with increased competition from unlicensed minicab companies, has resulted in emerging interdependencies between drivers and the companies they obtain work from, as well as among drivers, in some cases as small associated communities sharing such work. The aim of this chapter is to illustrate how work that has traditionally been both highly mobile and independent is increasingly conducted in contexts of emerging mutual interdependencies. The advent of mobile Information and Communication Technologies (mICT) plays a central role in establishing these interdependencies. The chapter documents dramatic changes where mICT almost serve the opposite purpose from that of traditional settings, where the purpose of mobilizing interaction through mobile phones and devices such as the BlackBerry is to allow flexible choices of context within existing boundaries of mutual interdependencies. Cabbies have always moved for work but can now negotiate interdependencies, a difference from office workers, who have always negotiated interdependencies, but can now physically move when mediated by the use of mobile technology.

400 years of mobile work

The first record of London carriages offering to transport people from ranks was in 1588, and the trading history of the black cab drivers in London dates back to 1662, when the service was established within the city of London. At that time, the term 'hackney carriages' was applied to horse-drawn carriages, which were later modernized as hansom cabs (*c.*1834) and were operated as vehicles for public hire. The term 'hackney carriage' was later applied to electric vehicles, which were put out of circulation around 1899 after several years of use. Horse-drawn carriages existed until 1947, when they were fully replaced by motor vehicles. See Table 10.1 for an overview of the developments from 1662 until today. The traditional iconic London black cab has been produced in a number of models and makes since the late 1940s. Currently the most common motor vehicles in service are the TXI and TX2 black cab models.

The aim of this chapter is to illustrate how mobile technologies combined with competitive market pressures in changing the ways London cab drivers work. This work has seen significant technological developments over a 400-year period. However, the recent changes with the introduction of mobile ICT are shaping interesting new work practices, in particular the emergence of possibilities for creating and enacting mutual interdependencies between cab drivers through this technology. These changes also have an impact on the everyday distribution and choice of work, as well as balancing the planned and unplanned events that occur during the shift at work.

The work of taxi drivers varies, like many other types of work, across countries and cities depending on traditions, specific circumstances, institutionalized habits and local regulations (Gambetta and Hamill, 2006). One strand of research has conducted studies in London, Paris, Bangkok and Tokyo of both

Table 10.1 Hackney cabs through history compiled from various sources[1]

Century	*Service description*
16th	1588 marks the first recording of hackney coaches when one of Sir Walter Raleigh's expedition members, Captain Baily, puts four coaches on the Strand for hire.
17th	Hackney carriage services start *c.*1662 as horse-drawn carriages. The term 'hackney' comes from the French term 'Hacquenée' meaning 'a horse for hire'.
18th	Horse carriage services developed further. Hackney carriages were heavy four-wheeled carriages drawn by two horses and having six seats. By the 1760s there were over 1,000 of these causing congestion.
19th	Introduction of the one-horse and two-wheeled Hansom Safety Cabriolet (cab) in 1834, featuring a clockwork mechanical taximeter. Brief period of electric hackney carriages, 'Hummingbirds', 1880–99 before the transition from hackney carriages to automobiles began with the invention of the internal combustion engine.
20th	1903 saw the first petrol-powered cab in London. Most horse-drawn carriages replaced by automobiles around the outbreak of the First World War. Vehicle evolution of the automobile led to the last horse carriage licence issued in 1947.
21st	Currently around 20,000 licensed taxis in London. Mobile technology increasingly supports taxi work.

Note
1 For example Georgano, 2000; Bobbit, 2002; Wikipedia and www.lvta.co.uk/history.htm.

the recruitment and management of taxi drivers (Skok, 2000, 2003, 2004; Skok and Baird, 2005; Skok and Kobayashi, 2005, 2007; Skok and Vikiniyadhanee, 2005). In some countries the use of computers in taxis has been a common sight for many years, whereas the London cabbies through their intense training have had less need for this.

In order to understand the emerging changes to London cab driver work we engaged in a study combining a series of qualitative interviews with video-recording of driver behaviour and interaction. The empirical background for this chapter is a nine-month study comprising 35 qualitative interviews with London black cab drivers, and 14 hours of videotaped observation of driver behavior using a cab-mounted video camera in order to gain an in-depth understanding of what the work involves.

Becoming a 'Knowledge' worker

The Public Carriage Office (PCO) is responsible for regulating black cab work. The PCO is also responsible for the regulation code for accreditation and practice of the service for the benefit and protection of customers. This organization is part of the Transport for London Authority, an institution that controls and regulates transport in the Greater London area.

The main task of black cab drivers is to collect and transport passengers from one physical point to another. Since 1851, certification as a black cab driver has involved passing a series of exams proving that 'The Knowledge' has been acquired. This differs from other cities in Europe and around the world. Candidate drivers make a number of appearances in front of a Public Carriage Office examiner. This examination is largely unchanged since its inception. The licence issued allows the driver to perform duties within a range of six miles from Charing Cross Station in London. Historically the Charing Cross Monument in front of Charing Cross Railway Station has been the point from which the land distance to London from any point in Britain is measured.

In the process of acquiring the Knowledge, candidate drivers are required to learn over 300 routes or 'runs' through London, as referenced in their Blue Book. This covers around 25,000 streets and destinations. The Knowledge itself is basically a process of elimination by examination and the volume of work involved is comparable to a three- to four-year university degree. The method used to evaluate candidate learning is by

> appearances, which are scheduled interviews based on an answer first-pass next basis, with reductions in the time between appearances as the driver candidate passes the scheduled tests.
>
> (PCO, 2007)

To facilitate the learning of the runs, most candidates work in small groups motorcycling around central London on small scooters. The purpose of these learning tours is to develop a feeling for the city's roads. Candidates learn to match the physical location of a place and how to get there, taking into account factors such as traffic, path shortcuts, alleys, time of day and congestion.

One of the first aims of a licensed black cab driver is to obtain enough funds to purchase his or her own vehicle. To this end, drivers commonly work overtime for their first two to three years as cabbies. According to superstition, the first journey a new cabbie makes must be free of charge, otherwise misfortune will fall upon the driver. The certified driver's green badge (See Figure 10.1) is the physical manifestation that he or she is qualified as a cabbie. During training most drivers establish long-term friendships and working relationships that serve as the basis for the formation of loose social groups that gather – as will be seen later in this chapter – at several locations and exchange information about where and when to work.

Driving alone – resting together

Work as a cabbie has traditionally for the most part been idiosyncratically defined and this is one of the features attracting many to study for a licence. Drivers are able to define their own timetables and schedules, based on their

Figure 10.1 A London taxi driver's green badge or license credential.

own particular circumstances. Optimizing income as a cabbie is very much about optimizing the place of work to meet fluctuating demands for services. As such, the London black cab driver is a prototypical example of how the choice of work context through physical mobility is the essential driver for profitability. The drivers argue that it normally takes several years to really learn how to make money as a cab driver through establishing rhythms of work and the necessary experience in where to be. So, while the process of accrediting drivers is conducted, through the PCO, within a highly formalized organizational setting, then the actual conduct of work has traditionally been highly independent work carried out by cab-owning drivers.

An integral part of most drivers' work is planning timed interruptions to their work where they will seek to join other drivers in order to share information about news, gossip or just discuss what is going on in the trade. A common place used by cabbies for their social meetings are taxi ranks, which are located around the city of London. These ranks are used for starting and ending daily shifts as well as places for physical breaks during their working day. There are approximately 500 taxi ranks in the London area. The purpose of taxi ranks is to provide the general public with a set location where they can hire a licensed taxi. The rank is the only situation where a taxi may ply for hire in a stationary position. Ranks are situated in locations where the public most need taxis, e.g. mainline London rail terminals, hotels and major shopping areas. Fifty per cent of ranks are within the two boroughs of Westminster, and Kensington and Chelsea. Over time, drivers acquire knowledge of the best times to approach a specific taxi rank. Sitting in a rank is a slow but secure way to obtain work. Some drivers specialize in working taxi ranks such as the ones located by London airports. Some drivers choose to work fixed long hours, others choose to vary the length of shifts depending on demand. Some drivers prefer to work nights, while others only work during the day in fixed shifts. This diversity of approaches to the individual organization of work is a reflection of their independence.

Cabbies also meet at one of the 13 remaining (of the original 61) cabmen's shelters maintained by a fund established in 1875. Most of these shelters are staffed by an attendant, who sells food and non-alcoholic drinks. The shelter has a kitchen in which the attendant cooks food for the cabbies. Shelters have seats and tables and can accommodate 10 to 13 drivers. Gambling, drinking and swearing are all strictly forbidden. The process of acquiring the badge through studying for the Knowledge is in itself also a source of community, in that cabbies who were candidates together also often meet up for lunch at certain cafes or shelters to exchange experiences and to socialize. The shelters, taxi ranks and other fix-points around London have traditionally functioned as places where cabbies can exchange information, in a similar way to that Orr (1996) documented when he related how photocopy repair engineers used coffee rooms as a place for knowledge transfer through telling 'war stories'. However, when drivers are away from the rank, they have traditionally not been able to access information about opportunities for jobs, etc. Radio circuits and later computer-cab systems allow drivers an alternative communication channel to exchange information about contextual situated work before only accessible by visiting a black cab rank.

Staying in touch with mICT

London black cab drivers mostly find jobs by virtue of their knowledge and experience without support from mICT. The strength of investing years of driving on a scooter around central London is that this knowledge is very difficult to compete against with technological means. One technology has, however, readily immersed itself in the cab and found a natural role. As passengers are increasingly talking on mobile phones, the drivers are experiencing a dramatic decline in friendly chatting with customers on the journey. The mobile phone has also successfully found its place in the cab cockpit and for some of the cabbies one mobile phone is not enough (see for example Figure 10.2).

One of the drivers studied had three mobile phones, one linked to the dispatch service, one being the main number to call and a third with very cheap call-rates to be used as the main phone to call from. The mobile phone has provided instant access to family members, friends and in particular to colleagues around London. It is quite common for drivers to keep in touch with colleagues on their mobile phone and exchange information about available jobs. One of the common reasons for choosing to become a cab driver is the working arrangement, which allows for family commitments and work to be flexibly coordinated. The mobile phone plays an important role in this coordination. In terms of work-related use, the mobile phone is frequently employed by cabbies to share information with each other about available jobs, traffic problems, etc.

Around 60 per cent of the 30,000 registered drivers choose to work entirely on their own. The reason for this high percentage is the attraction this type of work holds for individuals who prefer to manage their own work: the possibility to choose when to work and when to stop working is very attractive compared to other jobs normally taken by skilled workers in London, such as construction workers. In fact 80 per cent of the drivers interviewed had at some point worked in the construction industry prior to obtaining their badge. Cabbies often express their dislike for 'office work routine' or '8 to 5 work'. Overall they express dissatisfaction with the idea of getting ensnared by routine. Traditionally black cab work has been performed by white males between the ages of 20 and 60 from lower middle-class or working-class backgrounds. In particular, East Enders and Essex natives dominate the cabbie population, with women and ethnic minorities being under-represented.

Competitive pressures

Once a cabbie has gained his or her badge and has access to a properly licensed black cab, they can ply for work at the taxi ranks or when driving around London. The strict regulative practices for licensed cab drivers and their black cabs have always served as a barrier to entry for this kind of work. As a result, the supply of cabbies is generally lower than the demand, thus creating opportunities for the less strictly regulated minicab, or private hire, firms. These are only licensed to pick people up when prebooked in the minicab office (often advertising its services through a yellow flashing light), on the telephone or via a website. Minicab drivers are typically less well trained than hackney cabbies and the minicab vehicles do not have to pass the same stringent PCO tests. The minicab companies have in recent years managed to establish significant competition, for example, through innovative use of mICT, lower prices and market segmentation. Some companies, for example, offer newer and more comfortable vehicles with lower fares for specific routes such as those between airports and central London. Costs can also be reduced by adopting a *drivers for hire* business model with shared ownership of vehicles instead of the *one cabbie, one vehicle* business model preferred by licensed black cab drivers. The expansion of minicab companies has had an effect on the black cab service and, to address the issue of competition, new forms of organizing work have been put into practice. This effort centres on innovating black cab work through mICT. One example of direct competition between technology and tradition is the introduction of GPS to guide the driver. The commodification of GPS navigation systems has made it an affordable tool for minicab drivers with little in-depth knowledge of the streets of London and thereby increased both customer confidence in and experience of this service. The black cab driver's knowledge of London streets is being challenged by this technology. The annual competition between a cabbie with the Knowledge and a minicab driver with a GPS system has so far always ended with a victory for the black cab drivers, even if the GPS minicab drivers are getting closer and closer. Furthermore, some minicab companies have been able to challenge the official black cab drivers not only through employing drivers relying extensively on knowledge codified in GPS technology, but also by offering services based on complex portfolios of information services. This includes, for example, web-based booking of cars and SMS messages sent to customers, informing them when the car is on the way and when it has arrived. The minicab companies have been forced to be well organized through mICT as they do not have the licence to pick up customers on ranks or in the street. They rely entirely on mediated interaction with customers.

From self-referential to multi-referential mobile working

As explained above, traditional black cab work is conducted without the support of communication technology but merely as a combination of drivers choosing which rank to park the car at or which streets to drive along in order to locate customers. This traditional mode of black cab work is undergoing change, with mobile phones able to create ad hoc connections between drivers and with organizational technologies based on computer dispatch systems supporting centralized allocation of jobs to cabs. Apart from the traditional dispatcher model where customers call the dispatcher who then advertises the job to drivers, a more contemporary approach has been tested. Here, the centralized dispatch function is entirely automated and uses information about the position of the caller's mobile phone to locate the nearest black cab. When one has been found, the system automatically connects the customer to the driver's mobile phone and they arrange where and when to pick up the customer. This development has resulted in more complex cab cockpits as seen in Figure 10.2, which illustrates three examples of black cab interiors: the traditional cab, a computer cab with human dispatcher, and the automatic system connecting customers calling from a mobile phone with the nearest cab driver.

Suchman (1987) suggests distinguishing between planned and situated acts. Planning implies a degree of detail in the preparation for an intended act: for example, choices for start and end of shift, choice of days of work, or the primary zone of work. Situational acts are more a resource of action, in which the locations and opportunity determine the action. Idiosyncrasy, improvisation and knowledge are all useful tools when choices between planned and situated acts are complex (Suchman, 1987). The Knowledge and work experience as a driver is a resource for both planned and situated acts. Traditional black cab work is characterized by a high degree of idiosyncrasy as the lack of reliance on others allows for individual circumstances and preferences to guide the choices, for example, of where and when to start and end shifts. Also the varied and uncertain nature of obtaining work renders idiosyncratic choices a viable strategy for short-term decisions of either where to drive or which rank to join in order to get customers. Depending on their need to earn money, and how busy the roads are, cabbies might decide to take one or many breaks from driving, stopping at ranks or parking their cars.

Some cabbies choose to be at certain places in the city during their shift as experience tells them that this will lead to a higher probability of obtaining work. However, if no work is available, then they improvise using pre-defined strategies for generating jobs, either migrating to a new area or observing, and developing awareness of events in their vehicle's surroundings. This kind of improvisation can be characterized as 'rehearsed spontaneity' or 'planned serendipity' (Skok, 2003) and is deeply rooted in the culture of the black cab driver. It allows a wide range of situations to be handled in a consistently reliable way. Traditional black cab work has very limited contextual awareness in that the driver only has updated information about where he or she is. Sometimes too many drivers are concentrated in one area, such as a railway station where trains arrive at infrequent intervals and the number of passengers disembarking cannot be easily determined by observation of the surroundings. In this case, some drivers are forced to leave the station without passengers and migrate to other areas in search of hails, incurring extra operational costs.

Over the history of their business, black cab drivers have been loosely associated with cab networks that provide basic services such as shelter – the green cabmen's shelters mentioned above – car sharing schemes, etc. Mobile ICT also represent a shared service, but they present a paradox when applied to the taxi trade. Serious competition to a service that was previously dominated by black cab drivers has emerged, as GPS systems have become available to unlicensed

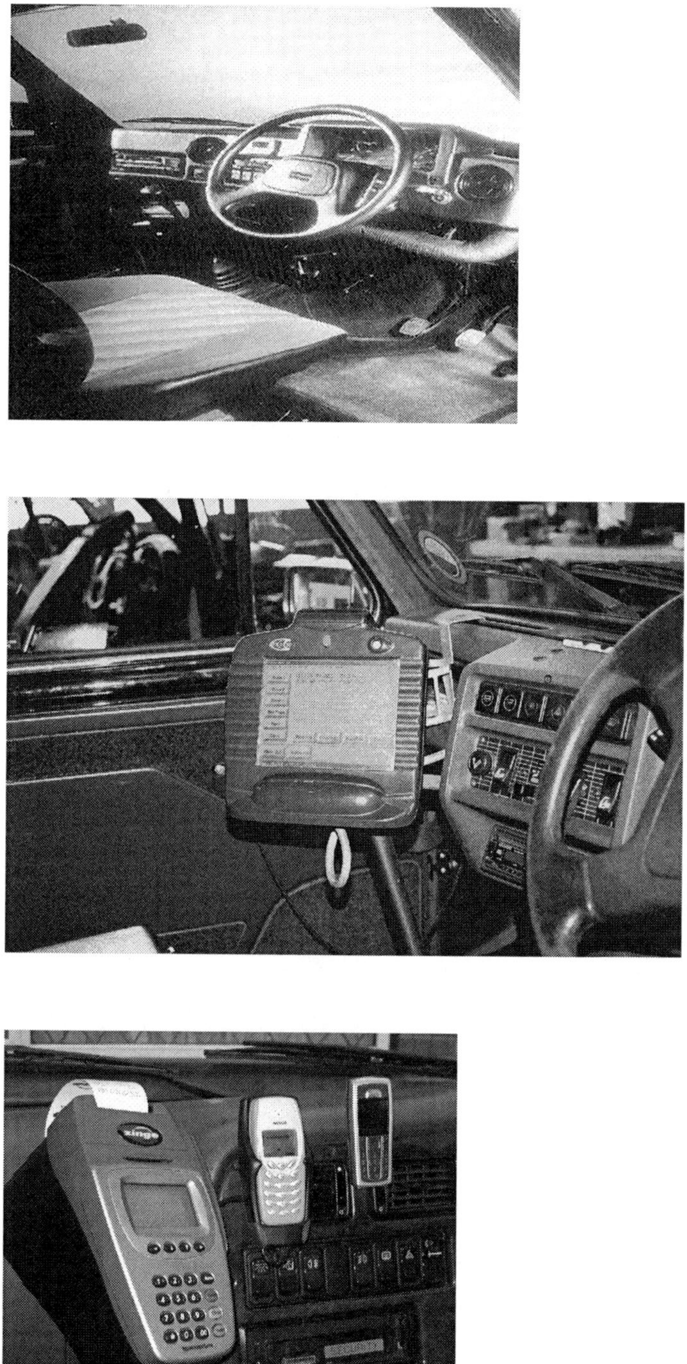

Figure 10.2 Three examples of variations in the installed mICT in hackney cabs.

minicab drivers. These systems enable drivers without the Knowledge to work the same routes as black cab drivers, and they are especially good when doing cyclical work such as picking up bankers from City Airport and driving into the Docklands financial district. In some cases, traditional sources of black cab work have dried up, as new working practices are defined by office workers able to work remotely or at home some days a week. Companies trying to reduce their office costs have also reshaped the demand for black cab services.

To their credit, black cab drivers have been able to rise to the challenge. Not only have the methods of managing the supply of black cabs benefited from the introduction of mICT, but the working methods employed by drivers have also changed. The most successful cab companies are thus able to compete in this environment with the aid of sophisticated technology. The main element in this shift is the application of technologies such as those illustrated in Figure 10.2 to support drivers in pooling resources and coordinating efforts. This can to some extent be conducted effectively in a highly decentralized manner through mobile phone connections in small and loosely coupled communities of drivers. However, joining one of the black cab networks relying on centralized coordination and dispatch enables drivers to gain access to jobs when they are difficult to find by driving around. The essence of this change one work, which is almost entirely self-referential towards forging mutual interdependencies amongst drivers and between drivers and the organizations dispatching jobs. For drivers there is a delicate relationship with centralized dispatch organizations, with membership requiring them to accept a certain number of jobs within set time limits. The subject of membership rules and advantages is conversely an important subject for discussion among drivers. One of the core issues is in that the principles applied in allocating jobs to drivers is either perceived as fair or suits their particular needs. The process is essentially one of establishing organization of mutually interdependent activities with competition as the catalyst and mICT as the primary tool.

There are several aspects to be taken into consideration when reflecting on what influences contextual choice in mobile work and the use of social resources is dynamic and unpredictable (Tamminen *et al.*, 2004). There are contextual and situated references of mobile work, where intermeshed situational and planned acts (Suchman, 2007) in event-led decisions have to be taken into account. In order to cope with uncertainty at work, mobile workers construct personal and group spaces within which to work, aided or supported by mICT. In some cases these choices made lead to the development of tensions that are dissolved as mobile workers learn to build social solutions to the use of the technology, such as mobile multitasking and awareness (Schmidt, 1998). The mode of interaction influences the progression from social practices to more systematic layouts in which the mICT becomes the central work tool as the collaborative complexity of coordinating interdependencies increases (Carstensen and Sørensen, 1996). Mobile workers deal with multitasking by taking either a self-referential or a multi-referential approach to their work.

Dynamic planning on mobile work

Supporting black cab work with mICT results in both planned efficiency gains as well as unplanned consequences. These gains were first noted by writers like Zuboff on shifts from manual to technological automation of skills (Zuboff, 1988) on static workers. The black cab drivers' use of GPS systems, for example, to speed up search for a location via a postcode search, can over time reduce the need for drivers to obtain the Knowledge or gently shift the practices towards increased reliance on the technology. Skok (2003) argues that the Knowledge constitutes a highly impressive and effective means for navigation. However, when not exercised on a daily basis, this may deteriorate and result in both poorer performance and greater reliance on technology. In the interviews, many drivers mentioned that they are aware that they use the Knowledge less, and regard losing it as a long-term risk.

In terms of the actual execution of work, the use of mICT does not give a clear picture of influence in the planned and unplanned acts for choices of work (Suchman, 2007). The mediation of the technology seems to shift choice from their experience to multiple other parameters. Work choices are influenced by the role of the technology in enhancing the awareness needed to do the job. Emergent practices also surge from this mediated technology, together with the already described human factors of working in isolation, physical location, perception of time and its distributed agent for fragmenting or compressing tasks (Elaluf-Calderwood and Sørensen, 2006). All these factors merge into a virtual raffle, where the chance of achieving economic success makes planning a very difficult task to describe, However, for the drivers, the exposure to these changes is not so complex and, as pointed out, some years of experience enhance the opportunities to improve success.

The characteristics of differences between traditional black cab work and the emerging interdependent working with mICT is summarized in Table 10.2.

Discussion

This analysis of the emerging changes to black cab work has highlighted the shift from self-referential to multi-referential working as mutual interdependencies are forged through mICT. The analysis also pointed out the increasingly complex interrelationship between embrained knowledge and knowledge encoded in supporting technologies (Blackler, 1995). The analysis clearly points towards the extensive Knowledge acting as an important barrier for ongoing technological development as it has for a long period been adapted to the specific requirements of black cab work (Hanseth, 2004). The new ways of organizing work, the migration to multi-referential working practices, rely on the dynamics of enabling workers to join when needed or wanted. However, these two concepts, needing and wanting, are relative and kept in a fragile balance. The more the driver relies on the system to locate jobs, the less he or she relies on their in-depth knowledge of where they need to position themselves to

Table 10.2 Contrasting the traditional arrangement of taxi work with modern practices utilizing mobile phones and other mICT

Traditional cab work	*mICT-supported work*
Low level of organizing mutual interdependencies	Dynamic organization of mutual interdependencies in order to pool resources for everyone's benefit
Regulated business with little competition	Regulated business with increasing competition from minicabs utilizing mICT to gain competitive advantages
Drivers engage in idiosyncratic behaviour	Idiosyncratic behaviour moderated according to constraints imposed by mICT and associated organizational procedures
High degree of physical mobility exclusively informed by immediate experiences	High degree of physical mobility both informed by immediate experiences and by access to remote information
Work highly situated	Enhanced situatedness of work through continuous negotiations of situations against information about alternatives
Self-referential work, with drivers deciding themselves where and when to work	Multi-referential work as mICT offers opportunities to coordinate efforts
The Knowledge is main information source	The Knowledge supported by mICT
Drivers alone when situated in context	Drivers engage in very light collaboration in context

maximize income. In each of these situations, their individual needs for work are being negotiated through technology in order to be balanced by the supply of work. Furthermore, the once weak ties and low levels of micro-coordination of other drivers' work are now actively encouraged and improved to maximize awareness. It is only through both competition and collaboration that there is a chance to achieve success in the search for work.

The various kinds of mICT deployed to support black cab drivers have contributed both to new working practices and to shifts in how this self-organized, loosely coupled community behaves in practice. Whereas before, taxi ranks and shelters were the primary venues for exchanging information and socializing, this is now in addition extensively conducted through mobile phone conversations. The personal networks established from the early days of studying for the Knowledge can now be further honed and extended through keeping in touch when driving around. At the same time, the companies offering cab drivers an organized means of getting jobs rely to some extent on these communities as the subject of exactly how each company's policy for membership influences the individual driver's freedom to choose their own working arrangements. The perceived fairness of company policies and the relative merit of individual systems are frequently discussed.

The mICT support two levels of increased mutual interdependencies between drivers and their surroundings. They allow for stricter organization of the allocation of jobs though formal rules for membership of a particular cab organization, and they support a more fluid exchange of information about available jobs, roadworks, traffic congestion, etc. This distinction displays some of the characteristics of Lanzara's (1983) distinction between formal and ephemeral organization, for example, in terms of the clear contractual boundaries and formal membership with external commitments of the cabbie signing up for one of the cab companies, as opposed to the fuzzy and fluid emergence of jobs through other drivers calling to notify, for example, that a train station needs many cabs to transport passengers.

In terms of Suchman's distinction between plans and situated action, we can observe some changes related to the use of mICT. For the individual driver, years of experience lead to the emergence of certain modes of operation or plans loosely guiding their behaviour. Although their work is clearly to a large extent still self-determined, they do not work without plans based on experience guiding their choices. However, mICT produce two forces, each slightly shifting the balance between individual plans and situated action. Joining a formal organization will result both in increased opportunities for situated action through available jobs announced by the system, as well as in increased need for planning as membership requires the fulfilment of a quota of jobs through the company. Cultivating an informal network of colleagues can help to increase the opportunities for situated action as opportunities emerge that would have previously remained unknown. However, this can be done within the context of established individual practices and planned behaviour guided by experience and personal preferences. The individual planned behaviour applied by drivers varies greatly with, for example, some cab drivers driving as long as there is work, and others rigorously adhering to the same working hours each day regardless of the availability of jobs.

It can be difficult to achieve trust in mICT and what it offers if nothing similar existed prior to the technological intervention. The path points towards establishing mICT with a high level of fairness and transparency in order to encourage adoption and satisfaction in its use by mobile workers. Trust is an important factor contributing to the adoption of a new technology at individual and organizational level (Kramer, 1999). Trust plays an important role when the competition – as in the case of the cabbies – is mediated by the technology. A strong consideration in joining a dynamic organization is whether the mobile worker can trust the system with a degree of fairness in allocating work. This is difficult to achieve, as the conventional organizational modes for work do not reflect well the working dynamics of these drivers, who are accustomed to flexibility and independence. The organizations that seem to support their work better are the ones that have a legal foundation fulfilling all established requirements, but still manage to work on a very loose basis, with voluntary rules of association. Few formal procedures need to be followed in the everyday work of making the management of these organizations focus on the collection and allocation of work. The driver's main concern is how the allocation of work reflects their beliefs of trust and fairness. However, control over whether to accept or reject a job is something the drivers are not eager to give up. Concerns are projected in terms of finding exchangeable or alternative use of the skills the workers have acquired over time. The mICT must, when serving to extend self-referential mobile workers' contextual awareness, create a system able to cope with both idiosyncratic behaviour and demands for operational efficiency.

The increased uptake of ICT in general and mICT in particular is of course not peculiar to the work of London cab drivers. Most roles in most organizations have seen computers and telecommunications systems develop into an essential part of everyday working practices. However, what this chapter has documented in other ways represents the opposite trend to traditional reporting on the role of mICT. The chapter has documented significant changes to highly independent work activities previously almost entirely defined by the discretion of the individual driver as a one-person business. This change has been mediated by various mICT and is driven by external competitive pressures as well as by technological opportunities. The new working practices developed through technological support have allowed organization to emerge in both formal contractual arrangements, allowing cab drivers to pool resources, as well as in increased informal communication via mobile phones, enabling each individual driver to be more responsive to changes in the environment. This development from extensive independence to increased interdependence through mICT constitutes in many ways the converse of the traditional discourse of mICT, that it promotes increased freedom to interact anytime, anywhere, with anyone. A traditional role of mICT has often been defined in terms of facilitating direct connections between information workers already highly interdependent upon the activities of their colleagues. The technology is seen as means of engaging in increased flexible working arrangements by breaking down spatial and temporal boundaries for interaction and thereby allowing mutual interdependencies to be negotiated outside the boundaries of the office. Here, the role of mICT is very much one of freeing up interaction restricted by organizational arrangements. In the black cab driver case it is to some extent also the mobilization of interaction beyond the ranks and shelters, but here between people who are highly independent of each other in their daily activities. Mobile ICT take up a more interesting role as the vehicle for organizing interdependencies and thereby offering pooled resources of not only one cab but a fleet of cabs all mobilized to serve customers. As such, we have documented the opposite development of freeing work traditionally closely defined by organizational arrangements, and instead shown an example of organization emerging as a result of mICT. This is not the only, or the first, example of such use of mICT to change the way cab drivers work. Computer-cab systems have existed across the globe for many years, but London black cab drivers are particularly interesting, as their work and skills for centuries have been highly independent and, since 1851, based on an extensive and rigorously tested knowledge of London. Black cab drivers have always been on the move and, with mobile technology, they can now finally engage in shaping mutual interdependencies.

What remains to be understood is how the organization of mobile information work can learn lessons from the cabbies in terms of utilizing the increased contextual flexibility as a business asset.

Note

1 For example Georgano (2000), Bobbit (2002), Wikipedia, and www.lvta.co.uk/history.htm.

References

Blackler, (1995) 'Knowledge, Knowledge Work and Organizations: An Overview and Interpretation', *Organization Studies,* 16: 1021–1046.

Bobbit, M. (2002) *Taxi: The Story of the London Taxi Cab*. London: Veloce Publishing PLC.

Carstensen, P. H. and Sørensen, C. (1996) 'From the Social to the Systematic: Mechanisms Supporting Coordination in Design, Computer Supported Cooperative Work', *Journal of Collaborative Computing*, 6, 4: 387–413.

Elaluf-Calderwood, S. and Sørensen, C. (2006) 'Organizational Agility with Mobile ICT? The Case of London Black Cab Work', in K. C. Desouza (ed.) *Agile Information Systems: Conceptualization, Construction, and Management*. London: Butterworth-Heinemann.

Gambetta, D. and Hamill, H. (2005) *Streetwise: How Taxi Drivers Establish Customers' Trustworthiness*. New York: Russell Sage Foundation.

Georgano, G. N. (2000) *The London Taxi*. Princes Risborough: Shire Publications Ltd.

Hanseth, O. (2004) 'Knowledge as Infrastructure', in C. Avgerou, C. Ciborra and F. Land (eds) *The Social Study of Information and Communication Technology: Innovation, Actors, and Contexts*. Oxford: Oxford University Press.

Kramer, R. M. (1999) 'Trust and Distrust in Organizations: Emerging Perspectives, Enduring Questions', *Annual Review of Psychology*, 50: 569–598.

Lanzara, G. (1983) 'Ephemeral Organizations in Extreme Environments: Emergence, Strategy, Extinction', *Journal of Management Studies*, 20, 5: 71–95.

Orr, J. E. (1996) *Talking about Machines: An Ethnography of a Modern Job*. Ithaca, NY: Cornell University Press.

PCO (2007) www.tfl.gov.uk/.

Schmidt, K. (1998) *Some Notes on Mutual Awareness*. COTCOS Awareness SIG Workshop, Paris: France.

Skok, W. (2000) 'Managing Knowledge within the London Taxi Cab Service', *Knowledge and Process Management*, 7, 4: 224–232.

Skok, W. (2003) 'Knowledge Management: New York City Taxi Cab Case Study', *Knowledge and Process Management*, 20, 2, 2: 127–135.

Skok, W. (2004) 'Knowledge Management: Taxis Parisiens Case Study', *Knowledge Management Research & Practice*, 2, 3: 147–154.

Skok, W. and Baird, S. (2005) 'Strategic Use of Emerging Technology in the Taxi Cab Industry', *Knowledge and Process Management*, 14: 295–306.

Skok, W. and Kobayashi, S. (2005) 'An International Taxicab Evaluation: Comparing Tokyo with London, New York and Paris', *Knowledge and Process Management*, 14, 2: 117–130.

Skok, W. and Kobayashi, S. (2007) 'Strategic Management of the Tokyo Taxi Cab Industry: An Exploratory Study', *Knowledge and Process Management*, 14, 1: 37–45.

Skok, W. and Vikiniyadhanee, J. (2005) 'International Management of Taxi Cab Operations: A Case Study in Bangkok', *Knowledge and Process Manag*ement, 12, 2: 140–149.

Suchman, L. A. (1987) *Plans and Situated Actions: The Problem of Human Machine Communication.* Cambridge: Cambridge University Press.

Suchman, L. A. (2007). *Human–Machine Reconfigurations – Plans and Situated Actions*, 2nd edition. Cambridge: Cambridge University Press.

Tamminen, S., Oulasvirta, A. *et al.* (2004) 'Understanding Mobile Contexts', *Personal and Ubiquitous Computing*, 8, 2: 135–143.

Zuboff, S. (1988) *In the Age of the Smart Machine: The Future of Work and Power.* New York: Basic Books.

11 Context matters

Un-ubiquitous use of mobile technologies by the police

Daniel Pica and Carsten Sørensen

Introduction

One of the persisting advances through information and communication technology is the ability to mediate interaction across distance. However, as argued by Olson and Olson (2000), distance matters, and situated human interaction cannot be replaced with mediated interaction without raising serious issues. Mobile Information and Communication Technologies (mICT) can be viewed as a means of reducing the effects of distance by allowing for the combination of situated interaction while maintaining connections to others. Mobile ICT have enabled new ways of working and of engaging with a variety of physical environments (Sørensen and Pica, 2003). The most dramatic conceptualization of this is probably the idea of anytime, anywhere interaction whereby there is friction-free access across boundaries to any person and any service at any time from wherever, through whatever medium may be desired (Kakihara and Sørensen, 2002). Katz and Aakhus (2002) define this as the paradigm of perpetual contact. To refer to this phenomenon Agre (2001) suggested the term 'always-on' world. The new workers, defined by the mobile technologies they use to produce and support encounters, are ones that seem independent of space and time (Kakihara and Sørensen, 2004). These tools enable a worker to contact and be contacted by customers, superiors and co-workers at any time and anywhere. In addition, mICT act as channels to access corporate information systems, and thus allow for the consultation and update of a number of databases relevant to the work. The logic of such a scenario is straightforward: the worker is now more independent, more efficient and better prepared to face customers' demands. As a dominant ideology, the concept of perpetual contact, as epitomized by mobile devices and ubiquitous computing, is linked to the ideal functioning of mICT in organizations. Despite the perceived benefits, such a scenario of ubiquity has given rise to a number of perplexities. Privacy advocates envision the rise of an Orwellian society. Technofiles view developments as uniquely positive. Technophobes see the technology as a means of controlling the population. These views represent extremes difficult to reconcile, as they depart from radically different ontological and epistemological levels. In a less theoretical and politically charged manner, and relying on a practice-based lens, this chapter investigates the idea of ubiquity and the paradigm of perpetual contact by investigating operational policing. A practice-based lens is concerned with the everyday use of technologies and generally subscribes to ethnographic methods. This lens is derived from the theory of *structuration* and sees technology and its users as co-constituting, in that technological use cannot be stabilized but is emergent in time, and the structures of technology emerge only through continuous and situated use (Orlikowski, 2000).

The police represent an epitome of an mICT-dependent organization (Manning, 2003) and can thus serve as an ideal case study to address the question of *how appropriate the paradigm of perpetual contact is to understanding mICT in work settings*. This question is explored through analysing extensive data from a seven-month observational study of operational police work within a British constabulary. In particular, this chapter presents and discusses Response Vehicle (RV) officers' use of mobile technologies. The result of this study proves the ambiguous role that mICT play in organizations and proposes a balanced view of these technologies based on the context in which they are introduced. Our study clearly confirms that context matters.

Section 2 presents the diverging views of the paradigm of perpetual contact in police organizations. Section 3 outlines the ethnographic research approach chosen. Section 4 characterizes the operational policing role of RV officers, and illustrates their work through a tale from the field. Section 5 analyses the findings of the study. Section 6 discusses the findings, and Section 7 concludes the chapter.

The paradigm of perpetual contact in police organizations

A substantial body of research has documented police work, and some of this has explored the implications of information and communication technologies for policing. Broadly, the perspectives on mICT for policing can be categorized into three: functionalist, symbolic, and critical, each holding differing ideas about the validity of the paradigm of perpetual contact.

The functionalist view sees mICT as an improver of police organizations and crime control. The rationale of this angle is rather straightforward. As technology provides more, better, and faster information about any given setting, the act of policing is improved. Furthermore, the view of the police is that of a collection of individuals working for a common goal, which can be rationalized and divided into distinct steps. This is reflected in the 1980s' and 1990s' emphasis on scientific policing and the *rational organization* (Manning, 1988). Chu (2001) offers an archetypical functional view that purposefully attacks any negative views of technology, stating: *Luddites in policing beware: the train is leaving the station* (p. 3). According to Chu (2001), properly implemented ICT will improve the ability of the police to attend emergency and non-emergency incidents, increase the personal safety of police officers, increase intelligence-gathering, and decrease costs of administration and operations. Speaking specifically of mobile technologies, Chu (2001) argues for the vision of *anytime and anywhere* access to data through ubiquitous, ever-present mobile artifacts.

In contrast, the symbolic view does not necessarily suggest that mICTs improve crime control. However, it does suggest that technologies help in transmitting an image of greater power. Thus, technology is viewed as a proxy or as embedded. This effect is seen ultimately as a deterrent for crime and an aid to controlling disorder. This specific view of the role of mobile technologies is reflected in the academic debate against perpetual contact and against the idea of *any time and anywhere* as an overarching concept. Lee and Sawyer (2002) illustrate how various physical arrangements of the working context relate to the temporal unfolding of activities. Wiberg and Ljungberg (2001) argue that the anytime, anywhere assumption is far from universal, as much work is conducted under spatial and temporal constraints. Cousins and Robey (2005) look at how mobile workers manage boundaries between personal and work activities and argue that the individual continuously engages in negotiating of these boundaries. The symbolic view suggests that ICT and mICT *have been constrained by the traditional structure of policing and by the traditional role of the officer* (Manning, 1992); and that they do not provide an instant solution to police problems but create contradictions and problems (Ackroyd, 1992). For instance, drawing on evaluation studies published in the 1970s and 1980s, Manning outlines the disappointing results of various technological innovations such as computer-aided dispatch systems, attempts to reduce response time, car locator and tracking systems, crime-mapping techniques, and management information systems, which failed to meet expectations and in some ways exacerbated original problems (Manning, 1988). He concludes that *such research as exists is often inconclusive or suggests that new technologies have less effect on police practices than their proponents predict or prefer* (Manning, 1992). Manning's conclusions demonstrate the failure of looking at technology from a purely functional viewpoint, and instead focus on the symbolic function of technology within police organization. Manning (1997) argues that ICT are not only employed because of their functions, but also because of the image of police that the police themselves seek to transmit, the assumption being that when a police force is effective in transmitting an image of efficient coordination, respect of laws and order tend to increase.

In stark contrast, the critical view suggests that mobile technologies do not improve crime control and police work – or, even if they do so in specific situations, their use brings fundamental new problems to the enforcement of order that could potentially displace and dwarf the good potential of the technology (BSSRS, 1985). In these critical studies (Marx, 1988b) presents a gloomy vision of the effect of new technologies in crime control:

> the new surveillance goes beyond merely invading privacy, as this term has conventionally been understood, to making irrelevant many of the constraints that protected privacy. Beyond the boundaries protected by custom and law, privacy has depended on certain inviolable physical, special and temporal barriers – varying from distance to darkness to doors to the right to remain silent … with much of the new technology, many of them cease to be barriers.

Marx (1988a) further argues that 'each small extension of surveillance can shift the balance between the liberties and rights of individual and the state … cumulatively they [new technologies] can change relationships and principles central to our form of government [democracy]'.

To sum up, from a functionalist view, technology is the *holy grail* of policing and as such the drive of current mainstream research is to see how to pump more technology into every aspect of police work. From a symbolic view, technology can be both positive and negative; thus a number of considerations have to be made before implementing it. From a critical view, technology, although affording efficiency, is an instrument of control that can have highly negative consequences on society and therefore its application should be dwarfed. This research tries to transcend the traditional schools and offer a practice-based understanding of the idea of ubiquity. The next section will discuss the research approach.

Research approach

The particular method employed to explore the interaction with mICT in the context of police officers in the UK uses a practice lens and aims at analysing and documenting the situations in which the interaction with mICT comes into being in a holistic and narrative manner. Thus, instead of following single occurrences of mICT usage or non-usage, these are analysed together through an ethnographic method. The data was collected with the aid of pen and paper, and a mobile phone camera. The use of pen and notepad mirrored how the officers themselves gathered evidence. This method for collecting and documenting data helped foster trust between the officers and the researchers because of the delicate nature of police work (Van Maanen, 1988). For instance, while having nothing to hide, most officers were concerned that more invasive data collection techniques would be engaged to evaluate their performance with technologies. Thus, the usage of more *invasive* data collection tools such as audio-visual recording equipment was considered inappropriate for the subjects of study.

A total of almost 120 hours of observation of Response Vehicle (RV) police officers attending emergency calls were conducted over seven months in 2003. The ethnographic study produced over 100 pages of handwritten observations and notes. These were transcribed into a digital format. Content analysis was performed with a focus on the mode of interaction with mobile technologies in operational policing and the situational characteristics and the nature of the incident attended. In addition, during the course of the analysis, interaction count was used to quantify the various variables in an attempt to provide a theoretical generalization of the findings (Silverman, 1993). The combination of generalization based on both qualitative and quantitative analysis of the extensive data from the observations and interviews established a sound basis for establishing generalizability of a theory within a setting (Lee and Baskerville, 2003). The time spent on each police activity was logged and used to create simple statistical data of the amount of time spent by each officer performing a certain activity in terms of the usage of mobile technologies. Some outliers were identified and excluded from the analysis. Member validation served as an integral part of the research method (Lincoln and Guba, 1985). Lincoln and Guba (1985) see member validation as *the most crucial technique for establishing credibility*, for it allows the presentation of a convincing account employing the perspective of the people under study. More specifically, member validation served to confirm the descriptions of the work context of officers and the results of interaction counts. Executive and detailed reports describing the job and giving recommendations for officers were presented to key representatives for the roles studied. This resulted in only minor changes, as the officers generally approved the descriptions as representative and precise.

In the ethnographic tradition, the presentation of the data is as important as the fieldwork itself (Atkinson, 1990; Clifford, 1988; Clifford and Marcus, 1986). The study of police work and mICT use is presented through impressionistic narrative. The specific value of narrative is to catapult the reader into a particular context (Van Maanen, 1988). The tale of operational policing aims to present an accurate and detailed account of how mobile technologies were used in the field, and also serve as evocative account of the emotional richness of the situations. The next section presents an ethnographic tale of a typical day of operational policing from the perspective of RV police. RV officers are often out of the office and travel great distances in highly idiosyncratic environments. They attend immediate-response incidents such as domestic abuse, burglaries and public fights. In addition, these officers have to keep public peace, escort prisoners, look for wanted people, patrol hot-spots, give a sense of police presence, appear in court as witnesses, collect information about crimes, as well as advise and calm down victims. The closest and most important link for RVs is the control room, with intense and ongoing communication though information streamed to the police car MDT[1] and through personal communication. The tale is written in the first person to maintain authenticity.

A tale of operational policing: a domestic with RV

It is three in the afternoon. John, a rapid Response Vehicle officer, has just arrived at work. He tells me that a domestic violence incident has just been reported through the radio. John takes a personal radio (PR) and equips all of us with stab-proof vests. We go to the vehicle parking lot just behind the police station. We enter the vehicle, turn on the blue light and sirens, and drive out of the station. Mary, his partner, is sitting in the passenger seat. While John is driving, Mary asks the control centre to send her more details about the domestic in progress. A few seconds later an alarm sound goes on and Mary is reading from a small computer screen the details of the incident (the MDT). While this goes on John asks Mary also to look at the log of the crime to see when the call took place and if there is a history at that address.

The log tells us that the woman is in a state of distress and that the control room has organized for an ambulance. Mary informs John that there is a history of domestic violence at the location and that attendance could be dangerous because the alleged perpetrator tried to attack an officer last time they were there. John tells Mary to ask for some backup if other cars are in the area. Through the PR Mary asks a controller for backup. We arrive at the incident. It is a council estate. Some people are standing outside. While John starts making his way to the apartment, Mary is calling the victim via the mobile phone to check the status of the incident. However, we get no answer. Mary keeps updating their position and situation via the radio. We decide to go up to the third floor. While we go up, we hear the ambulance arriving and we see another RV from the communal window. John repeatedly knocks on the door with no answer. The neighbour comes out of her house and claims she was the one making the call to the police and that the woman, as far as she knows, is inside the house. She adds that the guy who beat her has got away, probably by a car. While we speak with the woman, collecting some details, the door starts opening slowly. John and Mary ask me to step back. Mary, through the PR, updates the control room of the current situation. From the door, a woman emerges, clearly in a state of shock and bleeding from the nose. She is pregnant and keeps her hands on her stomach. Mary asks her if there is anyone in the house but she doesn't answer. Mary keeps asking the question with a calmer tone and finally she answers that there is nobody and that the perpetrator escaped using her car. John cautiously makes his way into the apartment to discover that there are no more people inside. Mary gives permission for paramedics to enter the scene after it is cleared of any possible danger.

While the woman is in the care of paramedics, Mary asks her some more questions regarding the domestic abuse. The woman gives the registration number of the vehicle taken by the perpetrator along with his name. Mary, using the PR, puts a warrant on the vehicle registration number. She then tells the woman that they are going to look for the man and asks how he was dressed and what he looks like. She keeps updating the control room with details as well as taking notes in her police-issued notebook. Mary told the woman to be calm and that the police would take care of it.

We go back downstairs. 'Let's go back to the police station!' says John. We enter the car and slowly drive back to the police station. The MDT is off and the radio keeps broadcasting crimes happening around us. Since it is almost the end of the shift, unlike other times where we would stand by and enquire about other incidents in the area through the MDT and radio, we go back to the office and document the incident we have just witnessed.

Analysis

On the basis of careful codification of the field notes, feedback from police officers, and theoretical reflections based on Kristoffersen and Ljungberg (2000), operational police work falls into five main activity types.

1 Standing by in car before incident;
2 driving to incident;

3 taking action at incident;
4 driving from incident; and
5 standing by in car after incident.

These five activity types are based on the fundamental distinctions between waiting in the car, travelling and attending the scene of the incident (see Table 11.1). This points to a generic cycle of operational policing where officers wait in the car before travelling to the incident. After this activity, they then engage in resolving the situation, after which they drive away from the scene and wait around in the car afterwards. The five activity types do not necessarily flow in the sequence prescribed and they can be recombined in a number of ways and the streamlined sequence presented is often broken up at various stages, for example, when RV officers return to the station after an incident. Furthermore, incidents may be engaged in parallel with less urgent ones being relegated in the background to re-emerge later.

Through the cycle of activities, the MDT and the radio alternate as the primary mobile technology. During the engagement in incidents, the personal radio takes precedence over any other mobile technology. In turn, RV officers use the personal radio as the primary mobile technology throughout the cycle of activities, due to the fact that engaging in incidents makes up for a third of the time RV officers spend in the field. This implies that, for the most part, the preferred modality of interaction with mobile technology becomes voice-driven. However, it must be noted that, compared to other activities, the use of radio is more discontinuous because the officer is busy dealing with people and solving problems. Only during stand-by time does the MDT take precedence over the radio, allegedly because it offers a quicker and more complete overview of the incidents in progress. It also allows the officers to negotiate the situation ahead through the passenger reading out the MDT report. Data-driven mobile technologies are seen as intrusive, for they require the attention of the officer to be distracted from the situation at hand. This seems not to be a problem when the officer has yet to choose which situation to tackle, but it becomes a problem once the officer engages in a situation. Hence, in many cases, mostly those surrounding engagement, both of these mobile technologies take a back seat in order to enable the officer to engage fully in a situation. Thus, there is no easy juggling between mobile technologies and dealing with emerging circumstances, as each is a detriment to the other.

Although Table 11.1 presents a ranking of the most frequently used mobile technologies, this does not necessarily imply that a mobile technology will be engaged. Their use depends on an officer's personal preferences. It is difficult to speculate on these personal preferences, and so this analysis shall not do so. However, it is important to understand that there is no uniformity of usage, nor is there any particular intensity of usage. What is interesting is that interaction with mICT tended to disappear around the engagement activity of police officers, and when it did not, it tended to take the form of voice interaction rather than data interaction.

Table 11.1 The estimated distribution of work activities between the five main operational policing activity types and the ranking of mobile technologies in use according to importance in each activity type, and summary of preferred modality of interaction

Role	*Mobile technology and interaction modality by activity type*					
	Technology ranking & virtual modality	*Standing by in car before incident*	*Driving to an incident*	*Taking action at the incident*	*Driving from the incident*	*Standing by in car after incident*
	Time (%)	15	25	34	13	13
RV	*Ranking*	1 MDT[1] 2 Radio 3 Mobile phone	1 Radio 2 MDT 3 Mobile phone	1 Radio 2 Mobile phone	1 Radio	1 Radio 2 Mobile phone 3 MDT
	Modality	Data and voice	Voice or data	Voice or none	Voice	Voice or data
	Interaction properties	Virtual interaction	Virtual and situated interaction	Situated interaction	Situated interaction	Virtual and situated interaction

The MDT, the radio and the mobile phone were generally engaged for different purposes, although some overlapping was present. The MDT generally served to overview a list of active incidents (21);[2] access detailed logs on an incident (17); glean information about the Police National Computer (PNC) vehicle and people (16). The radio was generally employed to update the geographical position of the officers (15); get further details on an incident log (11); update a crime log (10); update details regarding the PNC vehicle and people (9); and communicate with other officers (8). The mobile phone was used mainly to quickly and synthetically communicate with other officers (23).

There are a variety of reasons why these mobile technologies were used for such different purposes. The mobile phone was seen as an important means of horizontal communication, being the only technology that allowed flexible and direct communication with colleagues. Historically, the radio was the only way in which officers could communicate with each other. However, since the radio spectrum is a scarce, force-wide resource, such horizontal communication has to go through the control centre and thus could be overheard by everybody. Although in critical situations horizontal communication through the radio is a common occurrence, during less critical situations (which constitute the majority of those attended) it seems unreasonable for both officers and controllers to take up the radio spectrum.

The MDT mainly served to overview incidents, as it allows officers to work independently of the control room in gathering information. In addition, the MDT was considered extremely valuable and quick in checking registration plates and in identifying characteristics of wanted individuals. However, when it came to the actual involvement in an incident, the MDT was quickly replaced by the radio as a more interactive way of communicating. In fact, the radio served most of the needs of officers when engaging in an incident. The *discontinuity of use* of mICT can be observed in Table 11.1. The pattern that seems to emerge is that, when the police are attending an incident, they tend to stop using technology.

The use of mobile technologies throughout the cycle of activities can also be explained by the information requirements of officers. As the activity changes, so does the information that the officer considers important for the performance of the tasks at hand. In the stand-by activity, both roles are interested in a global view of the various incidents (see Table 11.2).

RV officers keep overviewing long lists of incidents and logs of a particular incident to decide which one takes precedence and requires their immediate attention. Once an incident has been selected or assigned to RV by the control room, information about how to reach a destination becomes of primary importance. However, this is not all that is required. RV officers consider a preliminary risk assessment of paramount importance. While the MDT is good at giving an overview of incidents and important clues about how to reach a destination, it does not seem to support risk assessment, at least in the form that officers would like. The radio, again, becomes the central mobile technology. Voice communication with the control room, other officers, and at times victims becomes the

Table 11.2 Information types required across the activity types

Role	*Operational police work distribution by activity type*				
	Standing by in car before incident	*Driving to an incident*	*Taking action at the incident*	*Driving from the incident*	*Standing by in car after incident*
RV	Active incidents queue. Wanted people.	Destination, risk assessment, vicinity and status of other vehicles.	Vicinity and status of other vehicles. Ongoing risk assessment. Positive ID of offender.	Status of custody. Active incidents queue.	Active incidents queue. Status of custody.

central concern of police officers. This is allegedly due to the fact that knowing that backup can be available promptly creates a sense of security. When engaging in an incident, three things are important to RV officers. First, they need to know the status and availability of other officers for backup purposes. Second, they value an ongoing risk assessment, where the information provided by the control room through the radio and MDT is balanced against the situation that the officer experiences – this implies the need for an open two-way communication channel between officers and control room, where the officer's impressions of the incident take primacy over the recorded information and are incorporated into the information system. Last, a positive identification of the offender, which furnishes information about the physical characteristics of the offender and crucially, the threat that the offender might pose to officers and other citizens, is of paramount importance.

Although the information system contains some of the information discussed, it is generally not enough for the officer to take effective action; the officer always resorts to situational clues contained in the environment of the incident. The environment of the incident is full of individuals, victims, offenders, witnesses and tensions. When the incident has been resolved (e.g. an arrest has been made, a person has been warned, a dispute has been resolved, etc.) the information requirements become similar to those of the stand-by activity. Again the officer looks at the active incident queue, and the incidents are scanned through the radio. This is also the case for the activity of standing by after an incident. This is due to the fact that the officer is expected to update the log via the MDT or radio on how the incident has been resolved. In addition, the level of risk becomes unimportant as that incident has been resolved.

It is then natural to assume that the dangers a police officer face, which are manifested in the uncertainty of any possible outcome that he or she may find, affect the usage of any mobile technology available. The analysis of Table 11.2 shows that police officers, when in an uncertain situation, attempt to convert any operational uncertainty into controlled risk. They manage the uncertainty through two main assets – the mobile technology that they have available, and their personal training and experience. In any given situation, it is natural that one asset should be given primacy over the other, and thus far the analysis has shown this to be the case. The *discontinuities of use* form as the police officers move from one activity to another, e.g. driving to an incident, engaging in it and standing by after it has been resolved. When driving to an incident, the police officer is engaged in intense use of mICT but, when the engagement takes place, the police officer becomes actively involved at the scene and most technologies rescind into the background. After the incident, the police officer again turns to mobile technology, creating a pattern of un-ubiquitous usage. This pattern reveals a natural process rather than a faulty one. For example, uncertainty is a factor that leads to an alteration of the need for information, as each police officer will prioritize his needs determined by his sense of safety to fit the specific context of the situation. Thus, while driving to an incident, it is in the officers' best interests to use the MDT to gain as much knowledge of the coming engagement as they can. They prioritize their information needs and concentrate on a balance between arriving at the engagement safely and assimilating all they can from the mobile technology before they get there. This balance is altered at the engagement, as the police officers have to gather information not available through mobile technology and, more importantly, they have to act. This obligation to act or engage at an incident necessitates that they stop using mICT. This shows that the context in which mobile technology is used, in this case the various police cycles, has a determinate influence on whether mICT are employed. The next session discusses these findings.

Discussion

Mobile ICT constitute essential aspects of police work. They fulfil and enhance both the need for instant communication, for timely coordination, and for command and control in police organizations, as they indeed may for other kinds of organizations. In addition, mICT fulfil the need for information that benefits police organizations both internally and externally – for inter- and intra-agency cooperation. The technologies are also of benefit to the daily operations of individual police officers by supporting transforming uncertainty into risk assessments. Overall, these benefits can be grouped together to form the idea of the fluidity of work activities and fuel the idea of ubiquity and the paradigm of perpetual contact. The more closely technology follows human activities, the more ubiquitous a role it is given, but equally, it easily recedes as it distracts, or gets in the way (Sørensen and Gibson, 2008).

Despite the perceived benefits of increased police efficiency through mICT, a few critical observations come to the fore as a result of the empirical analysis. First, the notion of mobility as it is envisioned in most contemporary research, emphasizes technical capabilities of devices, such as the promises of constant contact and monitoring, and the elimination of so-called *dead time*. This idea proves myopic for many failures occurring when technologies are introduced into any given organization. A subtle resistance may surface in the under-usage of the technical capabilities of the devices. This is most evident in the police context, whereby officers purposefully detach from the technology in order to act in the environment. Even more so, the symbolic creation of surveillance in the form of the possibility of continuous documenting dwarfs many applications of mobile technologies. This sort of consideration is not surprising, especially when compared to ones in key literature on information systems; the most obvious and powerful example is contained in both Orlikowski's (1993) and Ciborra's (1996) study of Lotus Notes where a mismatch between open and easy access to information and organizational roles and responsibilities led to user resistance.

The notion of mICT enhancing interaction any time and anywhere is one considering the use of technology from the outside in, thus emphasizing the technological opportunities offered. However, everyday life is lived from the inside out with decisions and actions being shaped by the totality of the lived experience (Ciborra, 2006). This is particularly important for police officers, who risk their health when they physically engage in extreme situations. Furthermore, the quality of the service is not necessarily better as technology improves because the quality is not directly linked to the speed at which information can be accessed. For instance, the resolution of certain circumstances in a police setting does not rest solely on the availability of information or on the speed at which it can be accessed, but requires a cautious understanding and interpretation of the social environment in which it unfolds. Hence, no matter how fluid a mobile technology becomes, it can nevertheless fail to provide the specific support that the police officer needs to resolve any given situation. This implies that mICT have to seamlessly disappear from the context of action to allow officers to make emergent decisions based on a contextual understanding. It is, for example, common that officers deliberately hesitate before engaging in a situation in order to be better prepared through obtaining information and reflecting upon the best course of action. More importantly, what appears as a productivity enhancement, instant access to organizational resources and people, is not always a positive improvement in policing. On the contrary, police officers need to be inoperative, because being inactive still gives a sense of presence in communities.

Finally, although there is a stress on protocol, which can be easily inscribed and enforced through mICT, in order for police officers to develop effective peacekeeping techniques, there is a need for experience – which relies more heavily on trial and error and on a constantly evolving relational nature with an environment. Mobile ICT thus far seem to promote an encounter-based nature with context rather than a relational nature. Thus, the idea of ubiquity and perpetual contact proves incomplete and cannot support a context in its wholeness. Those debating for the perpetual contact paradigm all have valid points to make. However, what seems to be most suited is a balanced view of mICT as not uniformly ubiquitous. Context very much matters.

Conclusion

The aim of this chapter has been to provide a detailed analysis of the role of mobile technologies for operational policing in order to test the validity of the paradigm of perpetual contact without resorting to politically charged theories such as those contained in the privacy literature. To accomplish this, an example of operational policing in the UK was presented. A detailed analysis of how operational police work was conducted with mobile technologies in the physical contexts of the police cars and the incidents was presented. This research contributes to the understanding of highly distributed work and use of mobile technologies within the police and puts forward the idea that efficiency in work is not necessarily linked to the ubiquity of technology. Rather, efficiency is achieved at times through the un-ubiquity of mICT. This opens the horizon to new venues for designing and implementing mICT through understanding the context of deployment. This will, however, remain a subject for further research.

Notes

1 The MDT is a system that provides information access and communication services to police officers via dashboard-mounted, touch-sensitive displays. It can be used to access the log of incidents, communicate with other cars, do background checks, and view and accept a list of incidents.
2 Numbers in parenthesis represent the number of occurrences witnessed.

References

Ackroyd, S. (1992). *New Technology and Practical Police Work: The Social Context of Technical Innovation*. Buckingham and Philadelphia, PA: Open University Press.
Agre, P. E. (2001). 'Welcome to the Always-on World', *IEEE Spectrum*: 12–13.
Atkinson, P. (1990). *The Ethnographic Imagination: Textual Construction of Reality*. London: Routledge.
BSSRS (1985). *TechnoCop: New Police Technologies*. London: Free Association Books.
Chu, J. (2001). *Law Enforcement Information Technology: A Managerial, Operational and Practitioner Guide*. New York: CRC Press.
Ciborra, C. (1996). *Groupware and Teamwork*. Chichester: John Wiley & Sons.
Ciborra, C. (2006). 'The Mind or the Heart? It Depends on the (Definition of) Situation', *Journal of Information Technology*, 21: 129–139.
Clifford, J. (1988). *The Predicament of Culture: Twentieth-Century Ethnography, Literature, and Art*. Cambridge, MA: Harvard University Press.
Clifford, J. and Marcus, G. E. (1986). *Writing Culture: The Poetics and Politics of Ethnography*. Berkeley: University of California Press.
Cousins, K. C. and Robey, D. (2005). 'Human Agency in a Wireless World: Patterns of Technology Use in Nomadic Computing Environments', *Information and Organization*, 15: 151–180.
Kakihara, M. and Sørensen, C. (2002). *Post-Modern Professionals: Work and Mobile Technology*. In Information Systems Research Seminar in Scandinavia (IRIS'25), Denmark: Copenhagen Business School.
Kakihara, M. and Sørensen, C. (2004). 'Practicing Mobile Professional Work: Tales of

Locational, Operational, and Interactional Mobility', *Journal of Policy, Regulation and Strategy for Telecommunication, Information and Media*, 6, 3: 180–187.

Katz, J. E. and Aakhus, M. (2002). *Perpetual Contact: Mobile Communication, Private Talk, Public Performance*. Cambridge: Cambridge University Press.

Kristoffersen, S. and Ljungberg, F. (2000). 'Mobility: From Stationary to Mobile Work', in K. Braa, C. Sørensen and B. Dahlbom (eds) *Planet Internet*. Lund, Sweden: Studentliteratur, pp. 41–64.

Lee, A. and Baskerville, R. (2003). 'Generalizing Generalizability in Information Systems Research', *Information Systems Research*, 14, 3: 221–243.

Lee, H. and Sawyer, S. (2002). 'Conceptualising Time and Space: Information Technology, Work and Organization', in F. Miralles and J. Valor (eds) *Twenty-third International Conference of Information Systems*, Barcelona.

Lincoln, Y. S. and Guba, E. G. (1985). *Naturalistic Inquiry*. Beverly Hills, CA: Sage Publications.

Manning, P. K. (1988). 'Symbolic Communication: Signifying Calls and the Police Response', MIT Press Series on Organization Studies, 9. Cambridge, MA: MIT Press.

Manning, P. K. (1992). *Organizational Communication: Communication and Social Order*. New York: A. de Gruyter.

Manning, P. K. (1997). *Police Work: The Social Organization of Policing*, 2nd edition. Prospect Heights, IL: Waveland Press.

Manning, P. K. (2003). *Policing Contingencies*. Chicago, IL: University of Chicago Press.

Marx, G. T. (1988a). 'The New Surveillance', in G. T. Marx (ed.) *Undercover: Police Surveillance in America*. Berkeley: University of California Press, pp. 206–233.

Marx, G. T. (1988b). *Undercover: Police Surveillance in America*. Berkeley and London: University of California Press.

Olson, G. M. and Olson, J. S. (2000). 'Distance Matters', *Human–Computer Interaction*, 15: 139–178.

Orlikowski, W. J. (1993). 'Learning from Notes: Organizational Issues in Groupware Implementation', *The Information Society*: 9.

Orlikowski, W. (2000). 'Using Technology and Constituting Structures: A Practice Lens for Studying Technology in Organizations', *Organization Science*, 11, 4: 404–428.

Silverman, D. (1993). *Interpreting Qualitative Data: Methods for Analysing Talk, Text and Interaction*. London: Sage Publications.

Sørensen, C. and Gibson, D. (2008). 'The Professional's Everyday Struggle to Ubiquitize Computers', in M. Elliott and K. L. Kraemer (eds) *Computerization Movements and Technology Diffusion: From Mainframes to Ubiquitous Computing*. Medford, NJ: Information Today.

Sørensen, C. and Pica, D. (2003). 'The Future is Rock-Fluid: On Mobile Work, Trust and Flexibility', *Whitepaper Orange and Enterprise*, LSE. 30 October. http://mobility.is.lse.ac.uk/.

Van Maanen, J. (1988). *Tales of the Field: On Writing Ethnography*. London: University of Chicago Press.

Part IV

Home–work dynamics

12 Mobile phones, spillover and the 'work–life balance'

Diannah Lowry and Megan Moskos

Introduction

Beginning with the first industrial revolution that saw sustained economic growth following the application of inorganic sources of power to the production process, understanding technological development and its application in the economic and social sphere has long been central to academic analysis. Continuing apace, the advent of new and more portable technological devices and their dissemination throughout the economic and social sphere has again come to invoke investigation and new theorizing about the role of new technologies and its association with 'flexible working' in the contemporary period.

The role of technology in work and employment has sparked much debate in recent years, with optimistic accounts on the one hand (citing, for example, flexibility, customer and supply chain responsiveness, and an increased skill premium for workers) and with more negative views on the other (for example, work intensification, a blurring of work/life boundaries and a deskilling of workers). Technology, however, is of course not homogeneous in its uses or in its impacts. A large body of literature exists and is still emerging regarding the consequences of new technologies such as the internet and email for workers and the organization of work (see, for example, Hipple and Kosanovisch, 2003; Castells, 1996; Bimber, 2000; Grigsby and Sanders, 1998; de Groen *et al.*, 1998; Shortcliffe, 1998). Literature, however, is only just starting to emerge which explores broad social issues surrounding the mobile phone (see, for example, Bianchi and Phillips, 2005; Grant and Kiesler, 2001; Hurme, 2005; Truch and Hulme, 2004; Ling and Haddon, 2001; Sørensen, 2004; Valcour and Hunter, 2005).

Given the ubiquitous nature of mobile telephony, it is surprising that the work mobile phone has largely avoided scholarly attention. Indeed it has been noted that there is a tendency to ignore the impact of mobile technologies on the 'unspectacular' or pedestrian aspects of everyday life, including everyday work–life (Geser, 2004: 4).

This chapter reports on findings arising from a wider study that sought to explore how the mobile phone has broadly impacted on the experience of work. The wider study examined how the workplace mobile phone may shape worker identity, and the economic impacts of workplace mobile phone usage related to knowledge and distributive supply chains (Lowry and Moskos, 2005). The focus of this chapter is on the workplace mobile phone and the work–life balance, the guiding question being 'how may the mobile phone shape the boundaries between public (work) and private domain, and how are these boundaries mediated or negotiated?'

While we are concerned with the 'impact' of the mobile phone in workplaces, the underlying assumption is not one of technological determinism, for while technology may have various effects, technology itself is shaped by social and economic forces (Mackenzie and Wajcman, 1985). As Sørensen and Pica (2003: 1) observe:

> The mobile revolution is not only a matter of people moving around carrying mobile technologies, it is also the radical mobilisation of interaction and socialisation processes. This mobilisation of interaction radically influences temporal, spatial and contextual aspects of interaction. We move from a strict linear sense of time defined by the clock towards 'social time' where the social context defines our sense of time. In turn, mobile technologies offer technical possibilities of rendering interaction with people and corporate information services fluid.

Work, non-work and mobile phones: spatial and temporal factors

The way the mobile phone may shape the work–life balance is likely to be complex, and requires some consideration of spatial and temporal factors. For example, Graham and Marvin (1996) discuss the geographical impact of telecommunications and the complexities and ambiguity surrounding telecommunications and cities. They present a parallel formulation of the respective functions of cities and telecommunications by proposing that, while both are technologies of communication, cities minimize space constraints to overcome time constraints while telecommunications do the reverse. Borrowing from Graham and Marvin (1996), we argue that, like cities, organizations are also technologies of communication. Organizations make communications easier by minimizing space constraints to overcome time constraints; while mobile phones make communication easier by minimizing time constraints to overcome space constraints. From this vantage, issues related to the impact of workplace mobile phones on the work–life balance can be identified.

Townsend (2000), focusing on the spatial and temporal factors associated with mobile phone usage, argues that mobile phones lead to an acceleration of the urban metabolism, as workers accomplish their work while commuting or moving from place to place within a city. Extending this analogy, we argue that the mobile phone potentially leads to acceleration in the *metabolism of the organization* since the distribution of organizationally owned mobile phones to workers may bind the efforts of workers even in non-work hours to the requirements of the organization, effectively 'tethering' them to firms. This has led many theorists (see, for example, Baldry, 2002; Bain *et al.*, 2002; Felstead *et al.*, 2002) to speculate that mobile phones facilitate already escalating levels of work intensification. Exacerbating this trend is the pervasiveness of 24/7 modes of operation in global industries and the associated expectation that workers will be available and 'at work' around the clock.

Mobile phones used for work purposes have thus been construed as a type of digital panopticon, where discipline based on time is replaced by continuous accessibility (Kopomaa, 2000). Different groups of workers may be prone to the 'tethering' effect in various degrees. For example in Australia, casual workers in particular, depending on the industry in which they are employed, are typically expected to be on call around the clock. Prior to the mobile phone, workers could choose not to answer their landline if they suspected that they would be called in to work with insufficient notice to make necessary arrangements, such as arrangements related to child-care or other home-care duties (Lowry, 2001). Many such workers are now given work-based mobile phones, and may be penalized if they do not answer each call. Penalties are likely to involve being taken off subsequent work rosters, leading to a loss of income. Since casual workers typically experience high levels of income insecurity, the 'tethering' effect of the mobile phone may be exacerbated for this group of already disadvantaged workers.

As noted above, there is scant literature on the impact of the mobile phone on the experience of work or the work–life balance and the literature that does exist is largely speculative and focuses on potentially negative aspects of the mobile phone and work–life balance. The study reported here attempts to empirically explore the range of ways in which the mobile phone may contribute to how the work–life balance is experienced.

The discussion is in four main sections. In the next section we unpack the concept, language and frameworks of the work–life balance, since the term is arguably not neutral. In doing this we develop a framework for our study. We then describe our empirical study and findings. Some of the determinants of how the work mobile phone shapes the boundaries between work and non-work are explored, and the findings of the study are then discussed.

Work–life balance: language, critique and frameworks

The language of the work–life balance

Before an examination of how technologies such as the mobile phone may shape the experience of the work–life balance, the concept of 'work–life balance' itself needs to be dissected, with each of the components of 'work', 'life' and 'balance' carefully defined. As Guest (2002) and Hyman and Summers (2004) correctly observe, in terms of both policy and practice, each of these constituent words present definitional and operational problems. For the purposes of this paper, 'work' is defined as paid employment. This does not get around the limbo time of commuting time, but we have for purposes of simplicity opted for subjective interpretations of work-related time that is not directly covered by paid employment. 'Life', in the context of 'work–life balance' and for purposes here is defined as non-work, that is time when we are not engaged in work. Life in this sense means free time, leisure time (in pursuit of specific autonomous activities or hobbies) and family time.

'Balance', in the context of 'work–life balance' is not so easily defined. It is a complex word with different grammatical meanings in different contexts. As a noun, it typically refers to a physical process such that there is an equal distribution of amount or weight. But as Guest (2002) correctly argues, this definition is inappropriate in the context of work–life balance, since both sides could be either very heavy or very light. Moreover, the type of balance between work and non-work sought by many people is not one with equal weighting. Hence a more appropriate interpretation of 'balance' may lie in its psychological meaning, for example, in the way we say people are balanced, grounded or stable. This notion is preferable since it implies human agency as well as the need to acknowledge that balance can have both a subjective meaning and measurement (Guest, 2002).

As a verb, balance also implies agency since it implies the potential to manage balance. But the verb 'balance' could be substituted with 'juggle', a term that implies an inherent tension. We are thus still left with the question, is it actually balance that is to be valued? Again, this serves to highlight the need to acknowledge the objective and subjective notion of the term in the context of work–life balance. Arguably for many people, the desired outcome is one of imbalance, when the chosen amount of say non-work time is dominant over work time. But the reverse preference must be acknowledged for those people whose sense of self and identity is synonymous with their work identity.

This has led some theorists (see, for example, Fleetwood, 2007; Shorthose, 2004) to question the very notion of the work–life balance and its political and rhetorical underpinnings. Shorthose (2004) acknowledges the worthiness of the sentiment of the concept, but provides a cogent argument that work–life balance campaigns are characterized by 'an inherent and limiting managerialism' (Shorthose, 2004: 2). He proposes that many work–life balance initiatives operate from the assumption that work involves creative interaction with the world. He also argues that work–life balance initiatives are mostly aimed at professionals, who are assumed to have a commitment and calling to their profession, rather than at manual and unskilled workers, who work long and unsocial hours due to a much more formal and overt economic compulsion:

> Work as it is currently organised is not the same as our capacity to interact creatively with the world. Such capacities have been largely hi-jacked.... Despite this, management have routinely advocated work as a route to creative self-expression. This is the offering up of the problem as the solution. The work-life balance campaign is but the latest in a long line of such managerial thinking.... It is evidence of the paucity of thinking about work–life balance to simply present a continuation of orthodoxy as a genuine change.
>
> (Shorthose, 2004: 2)

At the core of the critique lies the view that much of the work–life balance discourse reflects the individualism, achievement orientation and instrumental rationality that is fundamental to modern bureaucratic thought and action, and advances the idea that such discourse further entrenches people in the work–life *imbalance* that they are trying to escape.

We acknowledge these critiques of the notion of the work–life balance and understand their foundation. But we feel it necessary to continue to explore the concept, and particularly the role that mobile technologies play in it, in the hope that we may better understand the work–life balance not as a managerial initiative, but as a self-determined and subjective *experience*. Following Guest (2002) and others, we propose that the notion of human agency is fundamental to an understanding of what the work–life 'balance' means. A crucial component of this is how different people mediate and negotiate the boundaries between work and non-work, and the role of mobile technologies such as the mobile phone in this process. In order to inform this type of enquiry, we need some understanding of the relationship between work and non-work.

Traditional work–life balance models

Relevant literature reveals five main models which attempt to describe the relationship between work and non-work:

1 segmentation models;
2 spillover models;
3 compensation models;
4 instrumental models, and
5 conflict models.

Each of these models is outlined in Table 12.1.

As Guest (2002) notes, the five models are essentially descriptive, lacking inherent analyses of their causes or consequences. Thus, while each of the models as they stand may have some use *in situ*, their value together as an organizing framework is limited. A central problem with the models in Table 12.1 is that they are neither conceptually distinct nor mutually exclusive. For example, how are negative and positive spillover defined for different actors? And how is negative spillover different from the conflict model? There is also some blurring in the distinction between the compensation model, the instrumental model and even segmentation model. Guest (2002) cites an example of the instrumental model as when a worker may undertake long working hours and maximize earnings in order to purchase a home. It is unclear how this necessarily differs from the compensation model since any satisfaction associated with the purchase of the home compensates for the long working hours, and presumably the house purchase may ultimately facilitate segmentation between work and non-work. In any case, the compensation model, as defined in Table 12.1, is not appropriate for the purposes of our study, since we

Table 12.1 Five main models of the relationship between work and non-work

Model	*Proposition*
Segmentation	Work and non-work are two distinct domains of life that are lived separately with no influence on each other.
Spillover	Either world of work or non-work can influence the other in either a positive or a negative way.
Compensation	What may be lacking in one sphere, in terms of demands or satisfactions, can be made up in the other.
Instrumental	Activities in one sphere facilitate success in the other.
Conflict	With high levels of demand in all spheres of life, some difficult choices need to be made and conflicts may occur.

Source: Adapted from Zedeker and Mosier, 1990 and O'Driscoll, 1996, both cited in Guest, 2002: 258–259.

are looking to see what shapes the boundaries between work and non-work rather than a substantive consideration of satisfaction in either sphere. With this in mind, we concentrate in this paper on the model of spillover (positive and negative) as our organizing framework.

Crucial to the focus of our study here is our view that analyses of work–life 'balance' issues that neglect examination of possible underlying facilitating factors are somewhat flawed. In this chapter, we approach our thinking about work–life 'balance' by specifically exploring the role of the workplace mobile phone as a tool which may (or may not) shape the *boundary* and relationship between work and non-work domains, as well as shaping the consequences of that relationship.

Method

Due to the lack of systematic research exploring different workers' experiences of the use of mobile telephony in their work lives, an exploratory, qualitative research design was required, one which allowed us to access individual workers' experiences and interpretations of the mobile phone in their everyday work and non-work life. In line with our understanding of technology and its applications and uses being shaped by social and economic forces, it is our contention that we cannot understand the impact mobile phones have for work and non-work without understanding the meaning that the participants themselves attribute to their actions and views. The aim was to discover the concepts and meanings that are grounded in the social actors' 'everyday knowledge' and 'everyday sense-making procedures', as they related to the use of the mobile phone and work–life balance.

Participants were selected on the basis that they used a mobile phone in their work. Variation in age, sex, managerial level and occupation of participants was also sought. Potential participants from property, ICT, agriculture, retail and trades sectors were identified and contacted, constituting a purposive sample for the purposes of this study. We carried out a series of in-depth interviews with 15 women and men from a variety of occupations, sectors and age cohorts, and a focus group was also conducted with five people whose partners were identified as frequently using the mobile phone for work purposes both within and out of work hours. Interviews employed a common set of semi-structured, open-ended interview questions related to each participant's use of the mobile phone, their dislikes and likes regarding the mobile phone, and how they perceived it (if at all) to have changed their work and home life. These questions served as base-line questioning only. Detailed probing to further pursue all avenues of inquiry formed a major aspect of all interviews.

Findings

Overall, participants' narratives suggested that the work mobile phone shaped the work–life balance in both positive and negative ways. A major finding was that different groups of participants exhibited varying methods of using the mobile phone to manage the boundary between work and non-work. Occupation, age, status, bureaucratic structures and associated market forces were all found to differentiate participants' patterns of work mobile telephony and work–life balance experience.

Boundary management and spillover

As noted earlier, what constitutes positive or negative spillover may well be tied into individual differences and subjective experiences and values. For purposes here, we have defined positive spillover as the experience of accessibility to the non-work domain during work time. Negative spillover on the other hand, refers to the experience of work intruding into the non-work domain. We concede however, that the subjective experience of what constitutes either type of spillover *ideally* requires a more nuanced analysis. For example, there is (as we expected) some blurring between the subjective experience of spillover and instrumental agency. The following quote from one of the ICT managers is illustrative:

> When I'm at work I don't like to get personal calls. I feel that's an intrusion on my work time and I don't like people at work having too many calls either. I don't like to see people taking too many personal calls or too long calls. Personally I therefore have to model this and I don't like to receive calls from my family at work unless it is really important.
>
> (ICT Worker 3)

A further illustration of the subjective quality of positive or negative spillover was indirectly found in the narratives of the retail managers in our sample. They found that workers in their stores were receiving and making calls and text messages while 'at work', presumably giving the workers some positive spillover, but this was viewed as negative spillover by the managers:

> during induction all employees are informed that all mobile phones should be switched off and locked in a secure place.... Texting is probably used more than anything else. They'll have the phone in their pocket and they'll start texting. It's mainly the younger staff. There was an incident where a checkout operator answered the mobile phone while serving. That incident was about answering the phone. But texting is more common. They just don't want to leave their mobile phone. It's amazing.
>
> (Retail Manager 1)

Positive spillover

A number of workers indicated that the work mobile phone facilitated a desirable crossover of work and non-work spheres, viewing it as a tool by which they could maintain family contacts while at work:

> It can help your family life too because if you've got a work mobile then you don't need to be physically at work. So it's a balance as to whether the benefits outweigh the disadvantages.
>
> (ICT Worker 3)

> In a lot of ways it gives me more peace of mind, not so much at work but in my social and parenting life, knowing that no matter where I am I can still be contacted if an emergency occurs.
>
> (Tradesperson 2)

> My wife and I like the way we can contact each other during the day, even if one of us are at work. That's the beauty of it ... its part of life now isn't it. That's one of the pluses, its nice to be in touch with the family when you're working.
>
> (Property Manager 1)

Participants reporting positive spillover exhibited considerable agency in the way that they managed the boundaries between the two domains and also provided some insight into how the mobile phone facilitated positive spillover during commuting time. One participant took advantage of her commuting time to and from work to organize childcare arrangements and other caring duties associated with her two children. One of the ICT managers reported employing his hands-free device in his car during what he termed 'dead time' to organize meetings and obtain work briefings. Other workers simply reported that even if work spilled into non-work time, the nature of mobile telephony rendered the intrusion minimal:

> If you go back prior to mobile phones and you were on call, you had to hang around your own landline phone in case you got a call or you had a pager. If you had a pager it had to have reasonable range and you also had

> to be able to respond to that within a certain amount of time. For someone like me that likes the outdoors and likes walking it meant that I couldn't go too far from home because you had to be available to call back. With mobile phones it meant that you're no longer tied to home.
>
> (ICT Worker 3)

Negative spillover

Participants indicated a range of ways that the work mobile phone intruded into their non-work domains. In all cases of negative spillover, participants indicated a compulsion to answer calls, and reported feeling 'unable' to switch off.

> People think they can call you any time of the day or night. I got a call about a property at 7.17 this morning! And then you get calls at 11.30 at night, on Christmas Day. It's inconvenient and it's an invasion of your privacy, because you want to have your phone on.... I use my phone as an alarm clock, so I can't really turn it off. But now if I want to have a bit of a rest I'll turn it on to silent mode. It's very rare that I turn it off.
>
> (Property Manager 2)

Two of the participants in our sample had recently migrated to Australia from Peru, and one of them afforded the following cultural insight:

> Work life in developing countries is very different. In Peru I was on call for work 24 hours a day, everyone was. There is very high unemployment, so you need to keep your job, and you're just always on call. You can't disconnect from work, the mobile phone made you a slave. The only way to escape was to leave the country by going on holiday. I needed to explicitly say to my boss 'I will be out of mobile range, in fact I'm going to Alaska.'
>
> (ICT Worker 4)

One of the participants expressed the view that the ease of mobile telephony as a tool to solve problems quickly and associate actors in the bureaucratic web in decision-making processes, further facilitates negative spillover:

> Because you have a mobile phone you are accessible and people call me for obvious silly things. Co-workers and managers call you because somebody needs to make a decision about something – it doesn't matter what. Instead of making their decision themselves, they ask for help.
>
> (Focus Group Participant 3, previously an ICT Worker)

This intrusion of work into the home through the work mobile phone was not just felt by the workers. The narratives indicated some disharmony with other family members:

> My wife hates my work mobile phone. She hates it ringing, its intrusiveness. You sit down with your family and the kids always run after the phone and ask 'Dad, why are you on the phone again?'
>
> (Property Manager 4)

> I have a few hours in the afternoon with my six year old son, I won't turn my phone off but it depends what I'm doing whether I answer it. I usually do answer it, but it depends on my son's mood too. He sometimes gets grumpy, he'll say 'Are you on the phone again?'
>
> (Property Manager 3)

> I do feel very annoyed particularly about being disturbed at night as he often doesn't wake up and I would be the one who would say would you answer that phone and he would take a while to wake up and he would take the call in the bedroom then have to get up and start the computer up.
>
> (Focus Group Participant 2)

> It drives me nuts. He gets called on the mobile at all hours of the night, sometimes two or three times. He's fine at getting back to sleep, but I'm not. It's like his workplace has entered our bedroom. And the lack of sleep is affecting my own performance at work.
>
> (Focus Group Participant 1)

Discussion – the mobile phone and determinants of work–life balance

The study reported in this chapter attempted to address the way in which the mobile phone may shape the boundaries between public (work) and private domains, and how these boundaries may be negotiated and managed. Findings indicated that the way the mobile phone was managed to shape the boundary between work and non-work domain was related to age, occupation, status, economic context and the degree of instrumental agency that participants expressed in regard to their management of the work mobile phone.

Participants in our study were aged between 22 and 65. Overall, it was found that the older participants, particularly those aged in their 30s and over were more likely to use the mobile phone to actively manage the boundary between work and non-work, and were less likely to experience negative spillover. Younger workers had used the work mobile phone since entering the workforce, and appeared more prone to view it as a legitimized pervasive device of control, in terms of the way *it* controlled their activities rather than the other way around.

Negative spillover was reported more as an experience of the ICT workers and Property Managers, occupations with either elements of formality or bureaucracy in their work environments (the ICT workers), or elements of emotional labour in the course of their work (the property managers). Property management as an occupation is characterized by processes of emotional labour subject to the whims of buyers, vendors and managers, and remuneration is partly based on commission from sales. Against this backdrop, it is not surprising that the property managers in this study discussed feelings of being '*compelled*' to respond to their work mobile phones whenever or wherever they rang. The ICT workers in our sample were involved in 24/7 operations and part of a formal bureaucratic structure which extended globally. They were called at all hours of the night in the event of assembly-line stoppages or other outages caused by their ICT systems. In the event of a stoppage or outage, an elaborate reporting system and protocol were enforced, which meant that some of these workers would be contacted in their private domain during 'non-work' time.

The experience of participants in these occupations highlighted how the work mobile phone has facilitated a transgression of the boundary between public and private domains and shaped their experience of work more negatively. These workers' narratives revealed a sense of identity dissonance and anxiety, whereby front-stage work behaviour collided with their identities as 'mother', 'father', 'partner' or 'friend'. This identity dissonance in cases extended to the spouses of some of the workers ('it's like his workplace has entered our bedroom').

The participants in this study also consisted of management and non-management employees. The narratives revealed the consistent pattern of those respondents in management positions exhibiting more agency in the way they engaged the mobile phone to mediate the boundary between work and non-work. Non-management employees overall revealed a greater likelihood to experience negative spillover, indicating that mobile technologies may perpetuate existing power relations in work organization.

The propensity to allow the work mobile phone to transgress the work–home boundary is also likely to be embedded in the economic context of the type of work being performed. We do not wish here to rely on meta-narratives which serve to reduce complex phenomena 'away', however, some mention should be made of the economic context in which our participants' narratives are located. Market forces are dynamic and flexible, and our participants may have offered very different narratives at a different point in time and space (as evidenced by our example of cultural differences from our Peruvian participant). For example, the property managers in our study were operating in a static property market poised for a downturn, with a shortage of properties, sellers, vendors and general lack of property movement. If the market had been more active, they may have exhibited more agency in their management of the work–home interface, by virtue of being more financially secure. Along the same lines, there is currently a shortage of skilled trades workers in Australia. The high demand for trades workers places them in an atypical position of power, which helps to explain their greater control over work mobile phone usage, and their capacity to invoke and adhere to rules that served to seal the boundary between public and private domains.

Conclusion

Taken together, the findings generally support many of the assertions in the speculative literature mentioned earlier in this chapter. Workers most prone to reporting negative experiences of work–life imbalance were those who did not articulate conscious and definite strategies in the use of their work mobile phones to negotiate the boundaries between public and private domains. The difference in levels of agency between the participants can be partially explained by the nature of the work involved in the different occupations, the bureaucratic structures in which the occupations were embedded, and the economic context or market forces in which the occupations were situated.

The findings reveal a complex web of interrelated issues, reinforcing earlier work against the paradox of the work mobile phone being viewed as a 'double-edged' sword (Lowry and Moskos, 2005) and highlighting conceptual gaps in the work–life balance literature. There is a great need for more empirical research of this ubiquitous technology. We need to more fully understand the complex ways in which we interact and shape its form and usage, and how it in turn shapes and reshapes organizational communications and our related negotiations of the boundaries between work and non-work domains.

References

Bain, P., Watson, A., Mulvey, G., Taylor, P. and Gall, G. (2002). 'Taylorism, Targets and the Pursuit of Quantity and Quality by Call Centre Management', *New Technology, Work and Employment*, 17, 3: 170–184.

Baldry, C. (2002). 'Work Technologies and the Future of Work', *New Technology, Work and Employment*, 17, 3: 152–153.

Bianchi, A. and Phillips, J. (2005). 'Psychological Predictors of Problem Mobile Phone Use', *CyberPsychology and Behaviour*, 8, 1: 39–51.

Bimber, B. (2000). 'The Gender Gap on the Internet', *Social Sciences Quarterly*, 81: 133–160.

Castells, M. (1996–1998). *The Information Age: Economy, Society and Culture* (3 vols). Oxford: Blackwell.

De Groen, P., Barry, J. and Schaller, W. (1998). 'Applying World Wide Web Technology to the Study of Patients with Rare Diseases', *Annals of Internal Medicine*, 129, 2: 107–113

Felstead, A., Jewson, N. Phizacklea, A. and Walters, S. (2002). 'The Option to Work at Home: Another Privilege for the Favoured Few?', *New Technology, Work and Employment*, 17, 3: 204–223.

Fleetwood, S. (2007). 'Why Work–life Balance Now?', *International Journal of Human Resource Management*, 18, 3: 387–400.

Geser, H. (2004). 'Towards a Sociological Theory of the Mobile Phone', University of Zurich, available http://socio.ch/mobile/t_geser1.htm.

Graham, S. and Martin, S. (1996). *Telecommunications and the City: Electronic Spaces, Urban Places*. London: Routledge.

Grant, D. and Kiesler, S. (2001). 'Blurring the Boundaries: Cell Phones, Mobility and the Line between Work and Personal Life', in B. Brown, N. Green and R. Harper (eds)

Wireless World: Social and Interactional Aspects of the Mobile Age. London: Springer-Verlag, 121–132.
Grigsby, J. and Sanders, J. (1998). 'Telemedicine: Where It Is and Where It's Going', *Annals of Internal Medicine*, 129, 2: 123–127.
Guest, D. (2002). 'Perspectives on the Study of Work-life Balance', *Social Science Information*, 41, 2: 255–279.
Hipple, S. and Kosanovich, K. (2003). 'Computer and Internet Use at Work', *Monthly Labour Review*, 126, 2: 26–35.
Hurme, P. (2005). 'Mobile Communication and Work Practices in Knowledge-based Organizations', *Human Technology*, 1, 1: 101–108.
Hyman, J. and Summers, J. (2004). 'Lacking Balance? Work–life Employment Practices in the Modern Economy', *Personal* Review, 33, 4: 418–429.
Kopomaa, T. (2000). 'Speaking Mobile: The City in Your Pocket', available www.hut.fi/yksikot/YTK/julkaisu/mobile.html.
Ling, R. and Haddon, L. (2001). 'Mobile Telephony, Mobility and the Coordination of Everyday Life', paper presented at the Conference 'Machines that Become Us', Rutgers University, 18–19 April 2001.
Lowry, D. (2001). 'The Casual Management of Casual Work: Casual Workers' Perceptions of HRM Practices in the Highly Casualised Firm', *Asia Pacific Journal of Human Resources*, 39, 1: 42–62.
Lowry, D. and Moskos, M (2005). 'Hanging on the Mobile Phone: Experiencing Work and Spatial Flexibility', *National Institute of Labour Studies Working Paper* No. 153.
Mackenzie, D. and Wajcman, J. (1985). *The Social Shaping of Technology*. Milton Keynes: Open University Press.
Shortcliffe, E. (1998). 'Health Care and the Next Generation Internet', *Annals of Internal Medicine*, 129, 2: 138–140.
Shorthose, J. (2004). 'Like Summer and Good Sex? The Limitation of the Work–life Balance Campaign', *Capital & Class*, 82, 1: 1–9.
Sørensen, C. (2004). 'The Future Role of Trust in Work – The Key Success Factor for Mobile Productivity', *Tomorrow's Work Research Paper*, Microsoft.
Sørensen, C. and Pica, D. (2003). 'The Future Is Rock-Fluid: On Mobile Work, Trust and Flexibility', *Whitepaper Orange and Enterprises*, London School of Economics.
Townsend, A. (2000). 'Life in the Real-time City: Mobile Telephones and Urban Metabolism', *Journal of Urban Technology*, 7, 2: 85–104.
Truch, A. and Hulme, M. (2004). 'Exploring the Implications for Social Identity of the New Sociology of the Mobile Phone', paper presented at 'The Global and the Local in Mobile Communications: Places, Images, People and Connections' Conference, Budapest, 10–11 June 2004.
Valcour, P. and Hunter, L. (2005). 'Technology, Organizations and Work–life Integration', in E. Kossek and S. Lambert (eds) *Work and Life Integration: Organizational, Cultural and Individual Perspectives*. Mahwah, NJ: Lawrence Erlbaum Associates.

13 Freedom and flexibility with a ball and chain

Managers and their use of mobile phones

Keith Townsend and Lyn Batchelor

Mobile phones can be categorized as personal technologies and, as with other devices of this genre, increased use leads to increased reliance upon the technology (Sørensen and Pica 2003). In addition, mobile phones create opportunities, or at least the illusion of being able to create opportunities to 'make time' through providing temporal freedoms for the user. For example, if a someone is driving, they can return work-related calls, therefore making formerly 'non-productive' time 'productive' time. However, the 'rules' of mobile phone use in the workplace represent a relatively new issue for managers throughout the world. In addition, research on the use of mobile phones in the workplace is sparse. This chapter contributes to the small but growing literature in the area.

This chapter develops our knowledge of mobile phones in the workplace by presenting data from a survey of almost 100 managers. These managers are members of the Australian Timeshare and Holiday Ownership Council. Previous research has explored the manner in which employees engage their mobile phones to balance work and non-work life, and the tensions between managerial policy and employee actions (Townsend and Batchelor 2005; Batchelor and Townsend 2006). This research examines the manner in which managers of organizations deploy their mobile phones in order to be in multiple places at once. Mobile phone technology allows a dual presence (Fortunati 2002; Katz and Aakhus 2002). That is to say that these managers are able to maintain a presence in the workplace and outside the workplace at the same time, but the mobile phone also becomes somewhat of a 'ball and chain' that is constantly carried.

Introduction

Within organizations there are often tensions between 'what is' and 'what should be', especially in the area of policies and compliance. Organizational policies and operating procedures can often fit into the category of 'what should be' as they constitute a means by which the decision-makers in the company explain their expectations of employees'. Employees often find ways to resist managerially determined rules (Ackroyd and Thompson 1999) even when such rules are explicit. Complicating established methods of resistance are the rise of 'new' workplace issues. Managers are either unsure of how to formulate and implement new policy, or alternatively may fail to acknowledge the extent of the impact of the new issue on the workplace. Once issues are identified managers must strike an appropriate balance between instruction and information for employees, meeting legislative requirements for the protection of the organization while avoiding complicated rules that may hinder free-thinking and self-responsibility among employees.

Mobile phone use in the workplace is one such 'new' issue that managers of workplaces throughout the world are grappling with. This chapter presents a snapshot of mobile phone policies and their work use in one industry, before exploring the flipside of mobile phones, i.e. their non-work use at work. One of the concerns with the mobile phone that has been mentioned elsewhere (see, for example, Lowry and Moskos 2005; Townsend and Batchelor 2005) is the manner in which the technology is allowing a blurring of the work and non-work divide. This blurring is further investigated here to examine how managers are using their mobile phones, utilizing empirical material gathered from a survey of managers of member organizations of the Australian Timeshare and Holiday Ownership Council (ATHOC). The data presented in this chapter provide two aspects of the mobile phone and work nexus. We explore not only how work interrupts non-work life, but an under-researched area of mobile phone usage, how the mobile phone allows the non-work domain to intrude into the workplace.

The chapter is structured as follows. First, there is a review of literature examining the manner in which technology, particularly mobile phones, is being adopted and utilized. A discussion about our survey sampling and methodology follows, while the third section of this paper will examine the policy development aimed to guide employee use of mobile phones in these workplaces. Finally, this chapter will present the findings of this exploratory project, examining the manner in which ATHOC managers employ their mobile phones as a means of enabling a simultaneously dual presence between physical location and their virtual mobile phone presence. This dual presence means family and friends interrupting work time and work interrupting leisure time. Rather than merely confirming the intrusiveness of work into home life, the study reveals mobile phones facilitate leisure interrupting work life.

Adopting technology and developing policies

Mobile phones can be categorized as personal technologies and, as with other devices of this genre, increased use leads to increased reliance upon the technology (Sørensen and Pica 2003). Within organizations, mobile phones can provide employees (and indeed managers/employers) with a link between the world of the workplace and the 'outside world' or society at large. This is because a single device, the mobile phone, can serve both organizational and personal needs (Fortunati 2002; Srivastava 2005). People who require mobile phones for their work seem to have a greater level of flexibility and autonomy in completing their tasks and responsibilities while remaining reachable by others (Kakihara and Sørensen 2002, cited in Prasopoulou *et al.* 2006).

In addition, mobile phones create opportunities, or at least the illusion of being able to create opportunities to 'make time' through granting temporal freedoms for the user (Fortunati 2002). For example, if a user is driving, they can return work-related calls (using a hands-free device as legislated in many countries by 2006). In-car use means that waged or salaried time which would have formerly had limited or no productive value to the workplace can now be made productive. In the most positive sense, this dual presence means that the physical body can be separated from the place of activity (Katz and Aakhus 2002). Of course, this can be beneficial to the worker or manager, especially those with incentives and bonuses on offer, and these benefits then flow on to the organization. The use of travelling time in this way can also lead to work intensification for the employees. As the tool enabling driving time to become production time, this example demonstrates how mobile phones can be a double-edged sword for workers.

The example illustrates how the mobile phone has a spatial dimension where people can manage two 'spaces' at once, i.e. their actual location and the space at the other end of the telephone. These simultaneous physical and virtual spaces are described by Palen *et al.* (2001) and also Fortunati (2002). Interestingly, as Palen *et al.* explain, managing the physical and virtual spaces while on a mobile phone might require managing two 'identities'. Slightly varying an example given by Lowry and Moskos (2005), consider the example of a mobile employee dealing with an important but disgruntled customer face-to-face just as the employee's manager phones. The employee must balance the potentially competing identities of customer-service provider, as well as that of an employee assuring his/her manager that they are capable of handling their work. There are other situations that result in employees having to manage two spaces, one particularly difficult situation possibly occurring when employees deal with personal calls while at work.

Personal calls on mobile phones have a myriad of possible purposes and our empirical material shows how people find it difficult to ignore or terminate calls in potentially inappropriate circumstances. For example, consider the physical response that an employee may have if a new lover calls their mobile phone while at work. The softened tone of voice, perhaps giggling, or the dreamy look on the face of the person taking the call may be at odds with the physical presence expected in the workplace at that moment in time. Of course there is another view of the two spaces a mobile phone user is operating within. While any of the above-mentioned or other mobile phone calls are attended to, the people in the physical space are 'suspended', unable to engage with the phone user while the call takes place (Ling and Haddon 2001, cited in Lowry and Moskos 2005). It is not difficult for one to see many of the positive and negative aspects of mobile phone technology in the labour process.

There is an increasing awareness on the part of some organizations that mobile phones have become an ubiquitous item of personal technology rather than a tool supplied by an organization to facilitate job functions such as on-call service support. Mobile phones' constant workplace presence must be managed appropriately. Mobile phones support and in many cases underpin the anywhere, anytime flexibility that has become the mantra of many contemporary organizations (Wiberg and Ljungberg 2002). Yet, policies appear to be focusing on the important issue of health and safety rather than on more generalized questions of mobile phone usage (Badger and Mullan 2004). The need for focus on health and safety imperatives is well illustrated by an Australian Industrial Relations Commission case involving South Pacific Tyres, where personal use of a personal mobile phone resulted in a serious accident (briefly examined in Townsend and Batchelor 2005).

Developing policies for successful implementation at the workplace can be a difficult process. There is limited literature examining the process of organizations developing policies and manuals for their workforce. There is some reference to the filtering and reviewing of policies and manuals up and down organizational hierarchies (see for instance, Allan, in Gardner and Palmer 1998). However, more prevalent is a range of literature discussing the problematic nature of employee resistance when the policies are not implemented successfully (Thompson 2003; Townsend 2005; van den Broek 2002)

Methodology and the case

This research focuses on the views, policies and the use of mobile phones by managers of ATHOC member organizations. ATHOC is made up of representatives of the timeshare and holiday ownership industry. Established in 1994 to represent 'all interests involved in the timeshare industry, and to work towards national best practice' (ATHOC 2006), the council is primarily an industry association that also functions as a lobby group. The ATHOC website and newsletters promote the council as a 'a united voice for the timeshare and holiday ownership industry' (ATHOC 2006). Although this survey was administered in Australia, respondents included some of the large, multinational members. ATHOC represents organizations that range in size from large multinational interests such as Trendwest and Accor Premier Vacation Club to small, single-site, family operations. In total, ATHOC has more than 80 members, representing more than 120 holiday resorts and employing approximately, 4,000 people in Australia in occupations that vary from call centre staff to gardeners and food and beverage attendants.

The data was collected by self-directed questionnaire in two separate stages. The first stage was administered by the researchers at a conference, while the second involved the same questionnaire delivered as a web-based document. ATHOC was holding its Southern Region conference soon after the research instrument had been developed and pre-tested. The conference presented the rare opportunity to have 100 per cent response rate from the geographically disparate group. The paper-based questionnaire was completed individually by delegates at the conference, but within a room of more than 50 peers.

Despite it being a regional event, but due to the geographical spread of Australia, almost half of the council's members are members of this region. The respondents are 56 per cent male and 44 per cent female, with 5 per cent of the group under 30 years of age. The remainder of our sample show an even spread across five-year age divisions. Respondents are concentrated at higher levels of income, with 40 per cent of respondents earning more than $80,000AUD per year and almost half earning more than $70,000AUD, this skewed distribution is accounted for as the survey was restricted to managers. Some 61 per cent of respondents reported 'regular' office hours. More than three quarters of these 'regular' hours start between 8 and 9 am and finish between 5 and 6 pm.

After consultation with the council's general manager, the researchers employed an electronic survey to duplicate the study for the remaining ATHOC members. The survey website was developed using 'Surveymaker.com' and details were distributed to the remaining members via an ATHOC corporate email. The response rate for this form of distribution was substantially lower as the paper-based survey was somewhat of a 'captured audience'. It is interesting that the two groups show some significant differences in frequencies of the initial descriptive statistics run on some questions. However, as we are unable to ascertain whether these are methodological differences or differences between the 'regions' of ATHOC members, all data has been aggregated for this chapter. Hence, this chapter is analyzing the responses of 83 of a possible 130 respondents, or a 64 per cent response rate.

Existing mobile phone policies

We asked our respondent managers if they considered a mobile phone to be an essential tool in performing their work. More than 60 per cent indicated that this was the case. However, there is some discrepancy here between these figures and the 26.9 per cent of these organizations operating a general mobile phone policy. See Table 13.1.

From the written responses to our survey, we can establish the content of the general mobile phone policies in operation. These policies can be divided into four categories: mobile phones for work purpose only; short personal calls are

Table 13.1 Mobile phone as an essential tool and presence of policies (%)

		Mobile phones are an essential tool in my job		
		Yes	*No*	*Total*
My company has a policy regarding mobile phone use	Yes	26.9	28.9	55.8
	No	34.6	9.6	44.2
	Total	61.5	38.5	100

Note
N = 83.

accepted; only emergency personal calls allowed; and finally, no use of mobile phones in the workplace at all. Of the almost 60 per cent of respondents who say their company has a general mobile phone policy, almost three-quarters state that the phone is designated for work use only. Some 10 per cent of policies allow short personal calls, 5 per cent allow mobiles for emergency personal calls only; and 10 per cent of policies state that personal mobile phones cannot be employed in the workplace at all.

When respondents in workplaces without policies were asked if they thought that their organization should have general policies for mobile phone use, two-thirds responded in the affirmative. Written comments from respondents working with policies, almost completely categorized their existence as an issue of productivity. That is, managers found that employees were being distracted from their work by their own mobile phones and the mobile phones of others in the workplace.

This section of the chapter has discussed the presence of policies in the workplaces of ATHOC managers. We shift focus now to the managers' reports of what happens with mobile phone use inside and outside of the workplaces. These accounts do not always reflect the expectations of outlined organizational policies.

Mobile phone use in the workplace

It is possible that there is not a single workplace in a developed country that has not been affected in some manner (directly or indirectly) by technological developments in the last 20 years. Many organizations attempt to embrace the positive aspects of technological development while regulating or minimizing the negative effects of technology in the workplace. Some literature considers the inadvertent negative impacts of communication technology through interruption at work (Rennecker and Godwin 2005) and the empirical material from this research supports these findings. Our research has found that more than 80 per cent of ATHOC members receive mobile phone calls from non-work people every day.

Plant (2000) adopts the analogy of a mobile phone being like a calling bird – a ringing phone demands a response. The public, or workplace use of a mobile phone demands a response from all those within earshot. The response may be minimal, nevertheless the response is one of powerlessness. The only person who is able to deal with the incoming call is the owner or user of the telephone, even if it is an organization's phone, as the number may be attached not only to a particular person, but to a job function. Unlike the call of a landline, where an incoming phone call can annoy others in the workplace, the knowledge that a mobile phone may interrupt but can only be silenced by the phone owner, can be more irritating to those in workplaces (Plant 2000). In the ATHOC workplaces, of the 77 people who responded to the question 'in what ways do mobile phones have a negative impact in your workplace?', 64 responses (almost 85 per cent) referred directly to interruptions in the workplace, while many more referred indirectly to interruptions through commenting on the negative effect on productivity.

Townsend and Batchelor (2005) have investigated the blurring of the work and non-work divide with mobile phone use in small businesses. The employees in their study appeared to resort to mobile phones as somewhat of a 'band-aid' to cover imbalances in the employees' work and personal lives. We explore a similar issue here, but extend the research to find out just how often people's non-work lives exploit the mobile phone to encroach upon work time, and furthermore, to identify who is doing the interrupting.

Only 15 per cent of our ATHOC members suggest that they receive no non-work-related mobile phone calls or SMS (short message services or texts) a day. A total of 65 per cent of respondents report receiving fewer than three non-work-related interruptions per day (i.e. one or two). Like any level of time appropriation, small numbers can accumulate. Two interruptions a day seems minimal and acceptable but, if framed as the equivalent of 480 interruptions throughout a working year, some managers might reconsider the extent of the disruption. However, within our exploratory survey we are unable to extrapolate accurate lost time measures. Some 14 per cent of respondents suggest that they are interrupted more than three times a day – leading to potentially substantial blurring of the work and non-work divide.

Table 13.2 depicts the interruptions to the ATHOC managers while they are at work. Almost 60 per cent of respondents say that their partner is likely to call their mobile while they are at work. Just over 50 per cent say that their children will interrupt their work by calling their mobile phone. Not surprisingly, age is a substantial determinant here. For example, more than 50 per cent of those who report a 'partner' as the person interrupting are between 30 and 50. Around 45 per cent of those who report children interrupting are 45 and older; interrupting parents are concentrated in the younger ages; and interestingly, friends are substantially more likely to call the 55 and over at 13.5 per cent compared to a 3–5 per cent distribution in most other age groups. Males are twice as likely to be contacted by their partner; however, there is no sex differential for the interruptions from children and parents.

This section has developed a picture from the accounts of the use of mobile phones in the workplaces of ATHOC members. The next section explores what the ATHOC members have told us about the way that work-related mobile phone calls can overlap into their non-work lives.

Table 13.2 Who interrupts ATHOC managers on their mobile phones?

Who interrupts by calling mobile phone?	*Percentage*
Partner	57.7
Children	53.8
Friends	40.3
Parent	17.3

Mobile phone use outside the workplace

ATHOC managers detail the aspects of their mobile phone use which they find particularly important to their work. Our survey asked what tasks or possibilities mobile phones allow that would not have been possible five or ten years ago. Many of the responses revolved around the importance of mobile phones making communications more easy and effective. Wiberg and Ljungberg (2001) refer to the 'anywhere, anytime flexibility' that many contemporary organizations seek. Our data shows that ATHOC members overwhelmingly view this flexibility as a positive result of mobile phones. An operator of a small single-site operation states that their mobile phone is helpful because they 'can be in the garden weeding and answer the phone'. Other respondents refer to the ability to be informed and make decisions no matter where they are. In addition, many employees refer to their capacity to utilize 'dead time' such as driving to perform necessary tasks (one respondent even referred to making more productive use of time spent on the toilet!).

Almost 95 per cent of managers get mobile phone calls on work-related issues while they are engaged in non-work-related activities. More than three-quarters (77 per cent) of ATHOC managers say that they would 'seldom' or 'never' tell work-related callers that they could not talk if they were involved in a family activity. However, only 69 per cent of people would 'seldom' or 'never' tell non-work-related callers that they could not talk if they were at work. This suggests at least two possible arguments that are not mutually exclusive. Are more people privileging work over non-work time than vice versa? It seems that this is so. But are these people also using mobile as a 'band-aid' when their work and non-work lives collide? This also seems to be a reasonable proposition. We recognize that our data are exploratory and do not afford enough depth to fully explore the issue, and suggest more research be performed in this area.

Our study reveals who and what is being interrupted by work-related calls in non-work time among ATHOC managers. Almost 90 per cent of managers say that work has interrupted a family dinner, while 42.1 per cent say that a work-related call has interrupted their child's sporting activity. These figures are detailed in Table 13.3 (below), along with other activities that are interrupted.

Table 13.3 What gets interrupted by work-related calls?

Activity interrupted	*Percentage*
Family dinner	89.1
Child's sporting activity	42.1
Child's school activity	39.7
Weekend social activity (e.g. BBQ)	80.7
Relaxing night at home	85.5
Romantic dinner	42.1
Sexual activity	32.5

There is, of course, an obvious flaw in these data. We did not ask people if they 'attend their child's sporting events' or if they have 'romantic dinners'. Again, this research is yielding an early understanding of a new and growing phenomenon. Further research must be performed to understand the intricacies of patterns of mobile phone use as it relates to work.

When asked if a work-related mobile phone call interrupts sexual activity, almost one-third of our sample respond in the affirmative. While we did not ask employees where they were engaging in this sexual activity that gets interrupted, we assumed that it was going to be in 'non-work' time. Nevertheless, this question is worthy of further research as it illustrates the compelling nature of the call of a mobile phone ring tone, and the phenomenon of privileging a phone call over the physical present.

Discussion

The analysis of data presented here indicates a number of previously unexplored areas providing direction for further research and theoretical development in the area of the tensions between mobile phone use, workplace expectations and workplace behaviours. Supporting previous Australian research in the area (Batchelor and Townsend 2005; Lowry and Moskos 2005; Townsend and Batchelor 2005) mobile phone policies are presenting employees, from line managers to call centre workers, with a range of experiences in workplaces. There can be little doubt that mobile phones represent a technology that can assist many people in performing their work functions more efficiently and effectively. However, there is also little doubt that, in an age where people are finding it increasingly difficult to manage the balance between their work and non-work lives, they are resorting to mobile phones as a solution. Pocock (20035) refers to the 'work–life collision'. As with automobile incidents, injuries and ongoing worries arise from a work–life collision, and we suggest that many employees are using mobile phones as a band-aid on the cuts and abrasions left by this collision. Rather than structuring work in such a way as to avoid the collision, employees work around their absence from home by replacing personal presence with a phone call, or alternatively, the promise of answering a call.

Part of the problem giving rise to the need for a cure relates to the failure of many organizations to adequately develop workplace policies on mobile phones. Workplaces must have policies that at minimum address the use of mobile phones for health and safety-related issues. However, more comprehensive policies might include detailing for whom and when the use of mobile phones in the workplace is appropriate. Organizations may seek to sanction the use of mobiles provided that the phones remain on silent alert. A range of evidence suggests that companies who introduce work–life balance initiatives in other areas receive substantial benefits in a range of factors including worker commitment, and increased job satisfaction (Casper and Buffardi 2004; Konrad and Mangel 2000). Organizations that wish to promote work and non-work balance may use a mobile phone to do so; however, if this means employees are potentially working 24 hours a day seven days a week, then it is hardly promoting balance. Indeed, Porter and Kakabadsa (2006) report mobile technology uptake contributing to 'addictive' working practices. What might be preferred is if policy makers within organizations develop alternative policies that encourage employees to turn their personal mobile phones off. It is this phenomenon that is explained with one respondent who suggests: 'It provides freedom and flexibility for my work, but can be like a ball and chain.'

To a certain extent, this is exploratory research. The data presented here help us to understand the use of mobile phones and the presence of relevant policies within organizations that are members of ATHOC. From here, we can further explore the diverse membership of ATHOC to gain a snapshot of practices of non-managers in the service and hospitality industry as well as extend the survey to sister organizations and compare these results to studies of mechanized work.

Conclusion

The last decade has seen a dramatic shift in the manner in which people communicate in workplace and in non-work settings. Electronic mail, Voice (and video) Over Internet Protocol, SMS and MMS messages and mobile phones have made communication from person to person instant, whether the communicators are in adjoining rooms or on the other side of the world. This chapter has examined one aspect of this technology in particular, the mobile phone. What we have found through our survey data is that managers who use mobile phones consider the freedom and flexibility that this personal technology affords an asset in performing their work-related responsibilities. In addition, it allows them to maintain greater levels of contact with friends and family members while at work. However, there is a downside for these managers. The mobile phone can at times be like a 'ball and chain', a burden that they feel they simply cannot let go of for fear of missing an important call.

Managers, especially those in owner-operated businesses and small-to-medium enterprises, are often challenged when it comes to developing policies for the workplace. Policies relating to mobile phones are proving to be problematic for many organizations, including big businesses. This chapter has examined the extent to which some organizations have developed policies relating to the use of mobile phones in the timeshare and holiday accommodation industry. We have seen that there is a limited spread of policies in these organizations and, where policies exist, they may not be extensive or effective. Even where policies are in place, they only cover the commonly contested areas of workplace health and safety and areas brought to the fore by litigation. This study highlights the fact that organizations may need to look at productivity issues when considering how mobile phones, particularly personally owned phones, are deployed at work as it seems that much work time is being expended dealing with personal matters.

Mobile phones are likely to develop even more rapidly in the next decade, just as they have in the last. Companies must consult with their employees to develop policies that are effective to ensure employees, managers and the organization benefit from this developing technology as heavy-handed policies have the potential to create another avenue for employee resistance in the workplace.

References

Ackroyd, S. and Thompson, P. (1999). *Organisational Misbehaviour*. London: Sage Publications.

Australian Timeshare and Holiday Ownership Council Website (2006), accessed various dates, www.athoc.com.au.

Badger, F. and Mullan, B. (2004). 'Aggressive and Violent Incidents: Perceptions of Training and Support among Staff Caring for Older People and People with Head Injury', *Journal of Clinical Nursing* 13: 526–533.

Batchelor, L. and Townsend, K. (2006). 'A Modern Contested Terrain: Mobile Phones at Work', *Employment Relations Record*, 6(1): 64–72.

Casper, W. J. and Buffardi, L. C. (2004). 'Work–life Benefits and Job Pursuit Intentions: The Role of Anticipated Organizational Support', *Journal of Vocational Behavior*, 65: 391–410.

Fortunati, L. (2002). 'The Mobile Phone: Towards New Categories and Social Relations', *Information, Communication & Society*, 5, 4: 513–528.

Gardner, M. and Palmer, G. (1998). *Employment Relations*. South Melbourne: Macmillan Education.

Kakihara, M. and Sørensen, K. (2002). 'Post-Modern Professionals Work and Mobile Technology', paper presented at the 'New Ways of Working in IS: The 25th Information Systems Research Seminar in Scandinavia (IRIS25)', 10–13 August, Copenhagen Business School, Denmark.

Katz, J. and Aakhus, M. (2002). *Perpetual Contact: Mobile Communication, Private Talk, Public Performance*. Cambridge: Cambridge University Press.

Konrad, A. M. and Mangel, R. (2000). 'The Impact of Work–life Programs on Firm Productivity', *Strategic Management Journal*, 21: 1225–1237.

Ling, R. and Haddon, L. (2001). 'Mobile Telephony, Mobility and the Coordination of Everyday Life', paper presented at the Conference 'Machines that Become Us', Rutgers University.

Lowry, D. and Moskos, M. (2005). 'Hanging on the Mobile Phone: Experiencing Work and Spatial Flexibility', *Working Paper* 152, National Institute of Labour Studies, Flinders University, Adelaide.

Palen, L., Salzman, M. and Youngs, E. (2001). 'Discovery and Integration of Mobile Communications in Everyday Life', *Personal and Ubiquitous Computing Journal*, 5, 2: 109–122.

Plant, S. (2000). 'On the Mobile. The Effects of Mobile Telephones on Social and Individual Life', www.motorola.com/mot/documents/0,1028,333,00.pdf.

Pocock, B. (2003). *The Work-Life Collision*. Sydney: Federation Press.

Porter, G. and Kakabadse, N. (2006). 'HRM Perspectives on Addiction to Technology and Work', *Journal of Management Development*, 25, 6: 535–560.

Prasopoulou, E., Pouloudi, A. and Panteli, N. (2006). 'Enacting New Temporal Boundaries: The Role of Mobile Phones', *European Journal of Information Systems*, 15: 277–284.

Rennecker, J. and Godwin, L. (2005). 'Delays and interruptions: A Self-perpetuating

Paradox of Communication Technology Use', *Information and Organization*, 15: 247–266.

Sørensen, C. and Pica, D. (2003). 'The Future Is Rock-Fluid: On Mobile Work, Trust and Flexibility', *Whitepaper Orange and Enterprises*, London School of Economics.

Srivastava, L. (2005). 'Mobile *Phones and the Evolution of Social Behaviours', Behaviour* & Information Technology, 24, 2: 111–129.

Thompson, P. (2003). 'Fantasy Island: A Labour Process Critique of the "Age of Surveillance"', *Surveillance & Society*, 1, 2: 138–151.

Townsend, K. (2005). 'Electronic Surveillance and Cohesive Teams: Room for Resistance in an Australian Call Centre?', *New Technology, Work and Employment*, 20, 1: 47–58.

Townsend, K. and Batchelor, L. (2005). 'Managing Mobile Phones: A Work/Non-work Collision in Small Business', *New Technology, Work and Employment*, 20, 3: 259–267.

van den Broek, D. (2002). 'Monitoring and Surveillance in Call Centres: Some Responses from Australian Workers', *Labour and Industry*, 12, 3: 43–58.

Wiberg, M. and Ljungberg, F. (2001). 'Exploring the Vision of "Anytime, Anywhere" in the Context of Mobile Work', in Y. Malhotra and P. A. Hershey (eds) *Knowledge Management and Virtual Organizations*. Hershey, PA: Idea Group Publishing, pp. 157–169.

14 Travel, availability and work–life balance

Ann Bergman and Per Gustafson

Introduction

This chapter brings together two emerging research fields within the social sciences – on the one hand 'work–life balance' and possible conflicts between work obligations and family obligations (Guest, 2002; Hogarth *et al.*, 2000), and on the other hand, the individual and societal implications of human mobility (Cresswell, 2006; Sheller and Urry, 2006). The intersection of these two fields that we will explore here is work-related travel and availability.

Many workers today travel, not only to and from their ordinary workplaces but also as part of their jobs – to attend meetings, participate in conferences, visit customers, and so forth. Long-distance travel for work, and overnight travel in particular, may have considerable implications for working conditions as well as for family relations. Existing studies suggest that persons who frequently travel for work may often have a demanding work situation in other respects as well and, in addition, that such travel tends to be strongly segregated by gender and social class (Doyle and Nathan, 2001; Gustafson, 2006). For these reasons, work-related travel appears as a strategic factor to investigate from a work–life balance perspective.

The work–life balance discourse has been criticized, however, for being too individualistic and therefore 'genderblind', so that it fails to question structural inequalities between women and men (Smithson and Stokoe, 2005). In response to this criticism, the concept of 'availability' has been suggested as a complement to the notion of work–life balance (Bergman and Gardiner, 2007). More specifically the concept is advocated as an analytical tool for examining how individual actors are available in time and space, for work and for their families, and how their availability reflects an interplay between individual desires and choices on the one hand, and structural constraints and opportunities on the other. Arguably, temporal and spatial aspects of availability are central to work-related travel and its eventual impact on work–life balance.

In this chapter, we will examine to what extent women and men on different hierarchical levels in different kinds of work organizations regularly travel as part of their work, and how their travel is related to their availability, in time and space, for their work and for their families. For this purpose, we will use data from three different work organizations in Sweden – a paper and pulp mill, a regional bank and a university college (Bergman, 2004). Broadly speaking, these organizations provide empirical examples from the industrial sector (the mill), the service sector (the bank) and a more advanced knowledge sector (the university college). Previous research suggests that gender regimes, occupational structures and work-related travel may differ considerably between these different kinds of work organizations (Bergman, 2004; Gustafson, 2005). By exploring work-related travel in three different settings, and its relationship with other forms of availability at work and at home, we also hope to show the usefulness of 'availability' as an analytical concept within current discussions about mobility and work–life balance.

Availability

When discussing the relation between work and family, and the different demands connected with this relation, the concept of work–life balance is widely employed – both in political debates and in social science research. Research in this field has examined multiple social roles, the effects that work has on family life and vice versa, workers' behaviours and attitudes, health and stress (Hyman *et al.*, 2005). However, the concept has also been subject to criticism. Some critics claim that the notion of 'balance' is too harmonious and therefore fails to capture conflicts and unequal power relations; others argue that the concept is too subjective as it puts too much emphasis on individuals' thoughts and perceptions as to what constitutes a desirable balance (Guest, 2002; Lewis and Lewis, 1996; Smithson and Stokoe, 2005). In this chapter we will use the concept of availability as an attempt to get behind subjective perceptions of work–life balance, by focusing on working conditions and conditions in the family that can be more easily measured in objective terms.

Availability can be defined as the capacity of an actor or group to be available in response to the demands, needs or wishes of other persons or institutions (Bergman and Gardiner, 2007). This capacity can be a manifest activity under certain circumstances and to varying extent. The ways in which capacities of being available are manifest can be very different depending on the demands, the context and the actor who is available. Availability can further be conscious or unconscious, the object of free will or of force.

In working life there is a need for available labour power – individuals or groups that are available in different ways and to different extents, both in time and in space. Regarding the temporal dimension, most employment contracts specify that employees should work for a certain amount of time, and often at specified working hours. In addition, some organizations make demands that the labour force be available outside regular working hours, for example, by working overtime or by working unregulated hours. Spatial availability can take manifest forms as demands for presence at the workplace, or as work outside the ordinary workplace, for example, by employees taking work home or by being mobile (Bergman and Gardiner, 2007). Work-related travel, as will be discussed below, often involves demands for both temporal and spatial availability.

Different availability practices involve not only different demands from employers, but also different motives among employees. Psychological research on involvement in work sometimes makes a distinction between 'overworkers', who work long hours because of demands from the employer, and 'workaholics' (Peiperl and Jones, 2001) or 'willing workers' (Reeves, 2001), marked by an 'always on' work mentality (Reich, 2000), who may even work more than their employers demand. In practice, however, such a distinction seems quite difficult to make – there is often an interplay between individual motivation and demands from work organizations. Taken together, the tendencies described by Peiperl and Jones reflect a 'workification' or 'work-centredness', which is sometimes said to characterize contemporary society (Hochschild, 1997), or at least the upper segments of the labour market (Eriksson, 1998).

Family life, and the way it is organized, also require availability in time and space, so that different tasks can be done and needs satisfied. Several studies suggest that availability for family needs is more gendered – in terms of a traditional division of labour – than different forms of availability for work. While women tend to increase their labour market participation, it seems that they usually continue to take on the main responsibility at home (Duncan *et al.*, 2003; Grönlund, 2004). From a feminist perspective Jónasdottír (1994) argues that, in heterosexual relationships, women's availability for family duties liberates men from household and care work so that men can exercise their capacities to achieve more at work or in other spheres of life.

In this chapter, we understand work and family as multiple sources of demands for different forms of availability in time and space. We suggest that the investigation of these demands, and of their consequences in terms of concrete activities, is a valuable complement to the discussion surrounding work–life balance, where the focus is generally on how competing demands are perceived and how they can be managed.

Work-related travel

One specific working condition which often necessitates availability in time and space – beyond normal working hours and beyond one's normal workplace – is work-related travel. Travel for work has increased considerably during the past few decades (Doyle and Nathan, 2001; Swarbrooke and Horner, 2001), as improved technologies and infrastructures for travel have made possible more intense mobility, and as internationalization together with organizational developments in working life tend to require, more and more often, face-to-face encounters between persons working at geographically distant locations (DeFrank *et al.*, 2000; see also Faulconbridge and Beaverstock, in this volume). Several theorists have observed that 'mobilities' of various kinds play a crucial role in today's globalized world (e.g. Cresswell, 2006; Kaufmann, 2002; Larsen *et al.*, 2006), and frequent business travellers have often come to symbolize a mobile working life, where people are expected to work – or at least be able to work – anywhere and any time (Doyle and Nathan, 2001). Such 'hypermobile' workers are mainly to be found on high hierarchical levels and in knowledge-intensive organizations, where meetings and personal networks are important (Lassen, 2006; Urry, 2002).

Frequent work-related travel may have consequences for the travellers in several different ways. To begin with, travel-related stress is a growing problem among business travellers (Fisher and Stoneman, 1998; Striker *et al.*, 2000). Ivancevic *et al.* (2003) list numerous potential stressors – the need for planning and preparations before the journey, practical problems (delays, overbooking, lost luggage), cultural differences, isolation and concerns about personal health and safety during the journey, and accumulated work at the ordinary workplace when returning from the journey. In addition, some travellers tend to have very high workloads and/or long working hours during the journeys (DeFrank *et al.*, 2000; Doyle and Nathan, 2001). Travel may also have considerable consequences for family life. Long-distance and overnight travel means absence from home and family which may be stressful to the travellers as well as to their families, especially if the travellers have little influence over the timing and duration of their journeys, and/or if they have young children at home (Dimberg *et al.*, 2002; Espino *et al.*, 2002). Yet, at least some forms of travel also have important positive connotations, as they tend to be associated with self-fulfilment, an attractive mobile lifestyle, and career opportunities (Fisher and Stoneman, 1998; Lassen, 2006; Markham *et al.*, 1986).

Travel surveys suggest that work-related travel is not only hierarchically segregated, but also segregated by gender. Men generally travel more than women, and the travel activity is particularly low among mothers with young children (Gustafson, 2006; Presser and Hermsen, 1996). Indeed, travel has often been socially constructed as a predominantly male activity (Leed, 1991; Westwood *et al.*, 2000; Wolff, 1993), and it has been suggested that employers tend to regard women as less willing and less able than men to travel, partly because women are assumed to take on a greater responsibility for home and family (Davies, 2004). Taken together, these findings strongly suggest that work-related travel is an important issue to examine from the perspective of work–life balance and availability.

Case study organizations, data and methods

The data presented in this chapter originate from a quantitative comparative case study of three organizations – a paper and pulp mill, a bank and a university college – carried out in Sweden in 1997 (Bergman, 2004). The production processes of the three organizations can be described as more or less knowledge-intensive and more or less disconnected in time and space, with the university college at the upper end, the mill at the lower end, and the bank in between. Disconnected, in this context, means that work tasks need not be performed at a specific workplace or during specific working hours.

The mill is a traditional process industry producing paper and pulp, and was the largest of the three organizations with 1,277 employees. The mill was male-dominated with 20 per cent women and 80 per cent men. It was both horizontally and vertically segregated by gender. All hierarchical levels were dominated by men except the white-collar level. Table 14.1 shows the gender composition of occupations on different hierarchical levels, as defined by the Swedish Standard for Occupational Classification (SSYK), which ranks occupations in accordance with the qualifications they require. Employees in managerial occupations were almost exclusively male. Professional occupations, requiring for example engineering, economics or human resource management qualifications, were also male-dominated. White-collar occupations within administration, sales, logistics and support staff were female-dominated. Blue-collar occupations, comprising production, maintenance and transportation workers, were male-dominated and represented the vast majority of the total workforce.

The bank's core activity is the provision of financial services. Of its 317 employees, 68 per cent were women and 32 per cent men. Even though the women were a majority, the men were over-represented in managerial and professional positions, which mainly included legal and marketing occupations. Women dominated the white-collar occupations, which included financial advisers, clerks, cashiers and secretaries.

The university college's core activities are education and research. The organization had 598 employees and almost equal proportions of women and

Table 14.1 The organizations, hierarchical levels and sex ratios, (%)

	Men	*Women*	*N*
Mill			
Managers	97.2	2.8	106
Professionals	83.6	16.4	61
White-collar	35.4	64.6	147
Blue-collar	84.0	16.0	962
Total	79.5	20.5	1,276[1]
Bank			
Managers	64.3	35.7	42
Professionals	(75.0)[2]	(25.0)	12
White-collar	25.1	74.9	263
Total	32.2	67.8	317
University college			
Managers	71.6	28.4	67
Professionals	60.5	39.5	339
White-collar	25.0	75.0	192
Total	50.3	49.7	598

Notes

1 Information about hierarchical position was missing for one male employee.

2 Percentages in brackets are based on a total of less than 20 respondents (in this and the following tables).

men. Still there was an obvious vertical segregation with managerial and professional positions (mostly lecturers) dominated by men and white-collar occupations (administrators, secretaries, receptionists and clerks) by women.

The data-set was collected from two different sources: employer records and a questionnaire. Employer records were available for all employees, and provided individual-level data on occupation, hierarchical position, age, gender and working hours – data that were matched with the questionnaires. Questionnaires were sent to all employees, in total 2,192 (excluding hourly paid personnel), and returned by 1,664. The response rate was 74 per cent for the mill, 81 per cent for the bank and 77 per cent for the university college. The questionnaire provided data on work-related travel, overtime work and work performed at home, and also some additional demographic information.

The data collection and the research questions behind it did not primarily focus on work-related travel or work–life balance issues, and this implies some limitations with regard to analytical possibilities. Also, because of the case-study design, we have no intention of making empirical generalizations to the Swedish labour market as a whole. The findings presented should rather be regarded as empirical examples, which illustrate the analytical usefulness of the availability concept for research on work-related travel and work–life balance.

In the following empirical investigation, availability for work is operationalized in terms of temporal and spatial availability beyond a regular workplace and regular working hours (cf. Bergman and Gardiner, 2007). The main focus will be on work-related travel. In the questionnaire, the respondents were asked about travel for work (*resor i tjänsten*, which does not include travel to and from the ordinary workplace) that lasted longer than a normal working day. In most analyses we look in particular at those respondents who reported travelling every month or more often. Such travel demands availability both temporally and spatially, since it involves repeated absences from home for a longer time than is normally required for performing one's work tasks. With regard to temporal availability, i.e. the employees' availability in time beyond normal working hours, we also investigate to what extent the respondents worked overtime and whether or not their working hours were regulated. Spatial availability, i.e. availability outside the ordinary workplace, is also examined through a question about the extent to which respondents brought work home. As for availability for the family, we use two 'relative' measures – to what extent the respondents did more household work and/or less paid work than their partners. The latter analyses therefore include only respondents who were married or cohabiting.

The statistical elaboration in the following is restricted to descriptive cross-tabulations, examining work-related travel and other forms of availability among women and men, in different hierarchical positions and in the different work organizations. As relatively small proportions of the respondents travelled frequently for work, especially at the mill and the bank, more sophisticated multivariate analyses were not helpful.

Work-related travel in the three organizations

The following analyses are based on those 1,660 respondents who answered the question about work-related travel in the questionnaire. To begin with, as shown in Table 14.2, relatively few employees travelled regularly for work. In all, less than 10 per cent of the respondents reported travelling every month (or more often). However, the extent of work-related travel differed considerably between the three organizations under study. At the mill, the majority of the respondents – some 58 per cent – never travelled for work, and less than 6 per cent travelled every month. Work-related travel was more frequent at the bank, where over 60 per cent of the respondents reported some travel, and 11 per cent travelled at least once a month. At the university college, as many as 28 per cent of the respondents travelled every month and less than 12 per cent reported no work-related travel at all. Table 14.2 also shows that travel for work was strongly segregated by gender. In all three organizations, there was a clear and consistent pattern that men travelled more than women.

Previous studies of work-related travel suggest that travel activity may be segregated not only by gender, but also by social class. Table 14.3 provides a comparison between employees on different hierarchical levels in the three organizations. It displays the percentages of respondents within each occupational category who travelled for work every month or every week. In all three organizations, frequent work-related travel was more common among managers and professionals than among white-collar workers, and in the large group of blue-collar workers at the mill, only five respondents reported travelling every month. At the mill, the travel activity was a bit higher among professionals than among managers; at the bank and the university college, on the contrary, managers were far more likely than professionals to travel regularly for work. As many as 60 per cent of the employees at managerial level at the university

Table 14.2 Work-related travel and sex, (%)

	Every week	*Every month*	*Every six months*	*Less frequently*	*Never*	*N*
Mill						
Men	1.0	5.5	12.8	27.1	53.6	767
Women	0	3.4	7.3	14.5	74.9	179
Total	0.8	5.1	11.7	24.7	57.6	946
Bank						
Men	2.5	26.6	31.6	21.5	17.7	79
Women	0.6	2.8	14.7	33.9	48.0	177
Total	1.2	10.2	19.9	30.1	38.7	256
University college						
Men	8.7	26.9	43.8	14.9	5.8	208
Women	1.2	20.8	38.4	23.2	16.4	250
Total	4.6	23.6	40.8	19.4	11.6	458

Table 14.3 Work-related travel, sex and hierarchical level, percentage who travelled every month (or more often)

	Total	*Men*	*Women*
Mill			
Managers	23.6	24.4	(0)
Professionals	33.3	31.6	(42.9)
White-collar	15.0	29.3	5.1
Blue-collar	0.7	0.8	0
Bank			
Managers	29.7	47.8	(0)
Professionals	(9.1)	(12.5)	(0)
White-collar	8.2	22.9	3.8
University college			
Managers	60.3	60.0	(61.1)
Professionals	32.5	33.6	31.4
White-collar	8.1	14.7	6.1

college travelled every month. Taken together, Tables 14.2 and 14.3 indicate that the overall differences in travel activity between the three organizations were strongly related to the different occupational structures – the larger the proportion of managers and professionals in the workforce, the higher the travel activity.

Table 14.3 also provides separate analyses for men and women in different occupational categories. These analyses suggest that the gender difference in travel activity was primarily a reflection of the hierarchical gender segregation in the three organizations. To a great extent, men seem to travel more than women because they more often work as managers and professionals. At the bank, a considerable gender difference remains even when controlling for hierarchical level, whereas the remaining gender differences at the mill and the university college are less clear and less systematic. Yet those who travelled every month were few, especially among the women, so these analyses should be interpreted with some caution.

Travel and availability for work

The next step in our analysis will be to investigate to what extent work-related travel, and the demands that it brings along with regard to temporal and spatial availability for work, were associated with other demands for availability. To begin with, the questionnaire contained three helpful measures of the respondents' availability for work – overtime work, unregulated working time and work performed at home. The occurrence of these work-related demands varied considerably between the different organizations, as shown in Table 14.4 – all the three forms of availability for work were most common at the university college and occurred least frequently at the mill.

Table 14.4 Hierarchical level, sex and availability for work, (%)

	Total	*Managers*	*Professionals*	*White-collar*	*Blue-collar*	*Men*	*Women*
Work overtime every week							
Mill	14.0	56.2	57.8	33.0	3.4	13.9	14.5
Bank	51.0	83.8	(63.6)	44.4	–	58.2	47.7
University college	54.6	82.6	64.7	30.5	–	68.2	43.3
Work unregulated hours							
Mill	3.1	6.7	6.7	0	2.8	1.2	11.2
Bank	7.8	22.2	(20.0)	4.5	–	15.4	4.2
University college	57.1	72.7	77.8	17.9	–	77.5	40.2
Bring work home every week							
Mill	4.7	26.1	26.7	7.0	0.3	5.1	2.8
Bank	15.0	48.6	(27.3)	8.3	–	34.2	6.3
University college	59.4	75.0	76.6	23.0	–	75.1	45.9

Temporal and spatial availability for work was strongly related to the respondents' hierarchical positions. In all the three organizations, managers and professionals worked overtime, worked unregulated hours and brought work home more often than white-collar workers (and blue-collar workers at the mill). However, as the analysis of hierarchical positions in Table 14.4 shows, the different occupational structures of the three organizations did not fully explain the different demands for availability at the mill, the bank and the university college. Especially the two latter forms – unregulated working hours and bringing work home – were far more common at the university college even when controlling for hierarchical position. In several respects, availability for work also differed between men and women. At the bank and the university college men were more likely than women to report overtime work, unregulated working hours and bringing work home, whereas no such systematic pattern occurred at the mill. It should be noted, though, that some of the more fine-grained analyses are based on relatively few respondents, especially in the case of unregulated working hours.

Table 14.5 reports the proportions of frequent travellers (those who travelled every month or every week) and other respondents who were facing these three forms of demands concerning availability for work. The pattern is very clear indeed. In all three organizations, frequent travellers were more likely than other employees to work overtime, to work unregulated hours and to bring work home.

Analyses that examine the relationship between travel and other forms of availability for work while at the same time controlling for hierarchical positions or sex often have to rely on very small numbers of respondents (data not shown). Yet analyses that consider hierarchical positions indicate that frequent travellers, when compared with other respondents, systematically tended to face higher demands for availability for work in all the three respects investigated

Table 14.5 Travel and availability for work, (%)

	Travel every month or more often	*Travel less or not at all*
Work overtime every week		
Mill	67.9	10.7
Bank	75.9	47.8
University college	78.6	46.0
Work unregulated hours		
Mill	8.9	2.7
Bank	34.5	4.2
University college	76.0	49.5
Bring work home every week		
Mill	46.4	2.0
Bank	64.3	8.9
University college	80.8	51.0

here, even when comparisons were made between respondents on the same hierarchical level in each of the organizations. The differences related to travel activity were in fact surprisingly systematic.

As relatively few women were frequent travellers, separate analyses of women and men with regard to the intersection of travel and other forms of work-related availability were not useful for the mill and the bank. At the university college, frequent male travellers tended to face (other) demands for availability for work more often than frequent female travellers, but the differences were not dramatic.

In sum, then, the analyses presented thus far suggest that those who often travel as part of their job are more likely than other employees to face high demands regarding availability for work in other ways as well. They further suggest that such demands tend to be higher in more knowledge-intensive organizations and, in particular, among managers and professionals. Men also tended to meet higher demands than women, but in several cases this mainly seemed to reflect the fact that the gender composition differed considerably on different hierarchical levels and in the different organizations.

Travel and availability for the family

In this section we will investigate availability for the family, and the relationship between such forms of availability and work-related travel. The measures available here are 'relative' – the respondents were asked whether they did more or less household work than their partners, and more or less paid work than their partners (or if they and their partners did the same amount of household and paid work). Doing more household work and doing less paid work are taken as indicators of a relatively high degree of availability for the family, although the latter measure is somewhat indirect. Only those 1,353 respondents who were married or cohabiting are included in the analyses.

Table 14.6 suggests, to begin with, that employees at the bank and the university college were more available for their families than employees at the mill. In all three organizations, white-collar employees were more likely than all other occupational categories to be available for their families. The gender differences were also very clear, and strikingly similar between organizations. Most women but very few men reported doing more household work than their partners, and women were also far more likely than men to be in paid work for fewer hours than their partners (many of these women worked part-time). More detailed analyses, although partly relying on very small numbers, quite clearly suggest that gender is the predominant factor behind availability for the family. On all hierarchical levels women tended to be more available for their families than men, and the differences between the three organizations mainly reflect the different gender composition of their workforces.

The relationship between work-related travel and availability for the family is strongly negative (Table 14.7). In all three organizations, frequent travellers were less likely than other respondents to do more household work and less paid

Table 14.6 Hierarchical level, sex and availability for the family, (%)

	Total	*Managers*	*Professionals*	*White-collar*	*Blue-collar*	*Men*	*Women*
Do more household work than partner							
Mill	16.0	5.1	7.7	50.0	13.1	3.5	66.9
Bank	42.5	26.5	(18.2)	47.0	–	4.2	60.4
University college	37.5	15.9	30.7	56.7	–	4.9	63.7
Do less paid work than partner							
Mill	16.8	10.0	7.7	23.2	17.6	12.5	34.5
Bank	22.5	5.9	(9.1)	26.6	–	6.9	30.0
University college	22.0	6.7	14.1	40.5	–	6.2	34.7

Table 14.7 Travel and availability for the family, (%)

	Travel every month or more often	*Travel less or not at all*
Do more household work than partner		
Mill	2.1	16.9
Bank	7.1	47.5
University college	25.0	43.2
Do less paid work than partner		
Mill	8.5	17.3
Bank	3.6	25.3
University college	10.7	27.1

work than their partners. When comparing the different organizations, frequent travellers at the university college were somewhat more likely to be available for their families than frequent travellers at the mill and the bank.

With some minor exceptions (occurring in analyses based on very few respondents), these two patterns also appeared in separate analyses of women and men in the three organizations, as well as in analyses that control for hierarchical positions (data not shown). The strong gender division of household work was noteworthy, however. Among all the women in the study who travelled every month, over 50 per cent also reported doing more household work than their partners.

The analyses of work-related travel and availability for the family thus show that frequent travellers are less available for their families than other employees. Yet availability for the family – and household work in particular – was primarily determined by gender rather than by work-related factors. In all three organizations, on all hierarchical positions and even among frequent travellers, large proportions of women but extremely few men reported doing more household work than their partners. Similar patterns occurred when comparing household work and the different forms of availability for work (other than travel) examined in the previous section (data not shown). Most women who often worked overtime or who brought work with them home, and nearly half of those women who worked unregulated hours, also did the major part of the household work.

Discussion

The results presented here illustrate how work-related travel and other empirical indicators of availability for work and for the family may be used to examine the impact of a set of structural conditions on working life, on family life and on work–life balance. As argued above, structural conditions do not determine the outcome in each individual case – intentional (and at times unintentional) agency can make a difference. Yet our analyses very clearly show that gender, social class, occupational structure and the different production processes represented by the three organizations under study were associated with different patterns of availability among the employees, including different degrees of travel activity.

In all three organizations, men were more available for work than women, and women were available for their families to a much greater extent than men. Travel and other working conditions that implied availability beyond a regular workplace and regular working hours were generally more common among managers and professionals than on lower hierarchical levels. Availability for the family, on the other hand, was primarily segregated by gender – regardless of organization and hierarchical position, women tended to have the main responsibility for family and household work (cf. Duncan *et al.*, 2003; Grönlund, 2004).

To some extent, differences in availability patterns between the three organizations reflected the different gender composition of their workforces, but the organizations also differed in their occupational structures and in their production processes. Demands regarding travel and other forms of availability for work seemed to be higher in more knowledge-intensive organizations, in particular at the university college (cf. Lassen, 2006; Urry, 2002). Knowledge-intensive production was also, in the present study, related to a disconnection of work in time and space. Activities which need not be performed at a specific workplace and during specific working hours seem to produce higher demands for those kinds of availability that we have examined here. Indeed, different availability practices seemed to 'cluster', so that frequent travellers – usually managers and professionals – were often highly available for work in other ways as well. This pattern of occupational positions and availability for work, and its interrelationships, as demonstrated above, with gender and availability for the family, points at some important issues with regard to availability.

To begin with, availability for work as well as for the family is not only a matter of demands and obligations, but may also be desirable and rewarding. As discussed above, there are both 'workaholics', who want to work more than they have to, and 'overworkers', who have to work more than they want (Peiperl and Jones, 2001). Being highly available for work may imply either of these conditions, or a mix of individual strategies and ambitions on the one hand and structural demands, norms and expectations on the other. Although the term 'workaholic' may suggest addiction or disease, our analyses clearly show that those who are most available for work are also those who most often obtain the highest hierarchical positions in their organizations. Thus, 'being available' may be understood not only in terms of demands from the employer, but also as a resource and a competitive strategy among the employees – being available may be a way to accumulate organizational assets, to show loyalty towards the organization, and thus to promote one's occupational career.

It is at this point that the gender dimension, and the interplay between availability for work and availability for the family, become crucial. As women consistently, and apparently regardless of their occupational positions, tend to be more available than men for their families, their opportunities to use availability for work as a competitive strategy are limited. To put it sharply, it is not only the capacity of being available for work that is used as a competitive resource, but also the capacity of *not* being available for the family.

Rutherford (2001) argues that managers and professionals often develop a 'long hours culture', i.e. a high degree of temporal availability for work, and that such a culture tends to exclude women. Studies by Davies (2004), Markham *et al.* (1986) and Gustafson (2006) suggest that frequent work-related travel may have similar consequences. Because of women's and men's different degrees of availability for their families (and managers' expectations regarding such availability), women are probably less likely than men to be offered, and to take on, jobs that involve frequent travel – jobs that often imply high hierarchical positions. Several findings from the present study lend support to these suggestions: the clustering of travel and other forms of availability for work among some managers and professionals, the low proportions of women in these groups, and the high degree of availability for the family even among those (few) women who were highly available for work.

In discussions about work–life balance and family-friendly organizational policies, 'flexible' working hours and other arrangements that allow the employees to choose more freely when and where to work are often proposed in order to reduce conflicts between work and family obligations (Glass and Estes, 1997; Bond *et al.*, 2002). However, the availability perspective applied here suggests that such forms of 'flexibility' may be ambiguous. As long as women have the main responsibility at home, flexibility with regard to workplace and working hours may very well put more pressure on women to be available for family-related demands (Elvin-Nowak, 1998). For example, women who bring work home with them probably often – and more often than men – do so in order to be available for their children and for other family obligations (Nolan,

2002). Similarly, Bergström (2006) found that female travellers more often than male travellers avoided spending nights away from home when travelling, which implied travelling longer hours and longer distances in order to spend evenings, nights and early mornings with their families. Thus, the demands – and opportunities – for temporally and spatially disconnected work that characterize many knowledge-intensive organizations today may in fact have negative consequences for both work–life balance and gender equality.

In conclusion, mobile work and other temporal and spatial aspects of availability are important conditions in contemporary working life. They intersect with demands and desires regarding availability for the family, and are to a considerable extent structured by gender, social class, occupational positions and production processes. In this chapter we have argued, and tried to show, that the analytical concept of availability together with the empirical investigation of women's and men's availability patterns, in time and space, for their work and for their families, can advance current discussions about work–life balance.

The concept of availability highlights the demands that women and men meet, in working life and in family life, and how such demands influence their everyday activities. When we investigate patterns of availability, we see more clearly the conditions under which people attempt to obtain a satisfactory work–life balance, and we see how these conditions differ between different categories of employees. Moreover, we see that availability at home and availability at work do not constitute a zero-sum game – which the notion of 'work–life balance' may suggest – but something far more complex, in particular when regarded from a gender perspective.

Yet the full potential of the availability concept was not exploited here, partly because the concept was not fully developed when the data were collected. In particular, the measures of availability for the family were weak in comparison with the work-related measures. Further research in this field will need to consider and examine more systematically in what ways women and men may be available, in time and space, for work and for their families. In addition, the focus here has been on the effects of a set of structural factors on availability. The discussion above about availability as a resource clearly shows that further research will also need to address the interplay between such structural factors and individual capacities and strategies with regard to availability in different spheres of life.

Acknowledgements

Ann Bergman is grateful to Steve Fleetwood and his colleagues at the Institute for Advanced Studies, Lancaster University, for giving fruitful comments on her work on availability and work–life balance, and for providing an inspiring research environment during her stay in Lancaster. Per Gustafson's participation as co-author of this chapter was part of a research project on work-related travel, conducted together with colleagues from Göteborg University, and financed by the Swedish Council for Working Life and Social Research. Jan Ch. Karlsson provided very helpful comments on a late draft of the text.

References

Bergman, A. (2004) Segregerad integrering: Mönster av könssegregering i arbetslivet. Karlstad University Studies 2004: 46. Karlstad: Karlstad University.

Bergman, A. and Gardiner, J. (2007) 'Employee Availability for Work and Family: Three Swedish Case Studies', *Employee Relations*, 29: 400–14.

Bergström, G. (2006) 'Ett arbetsliv i rörelse: Resandets positiva och negativa potential för arbetstillfredsställelsen bland resande säljare', *Arbetsmarknad & Arbetsliv*, 12: 147–60.

Bond, S., Hyman, J., Summers, J. and Wise, S. (2002) *Family-Friendly Working? Putting Policy into Practice*. York: Joseph Rowntree Foundation/Policy Press.

Cresswell, T. (2006) *On the Move: Mobility in the Modern Western World*. New York: Routledge.

Davies, K. (2004) 'Mobile Solutions – Mobile Lives? Or the Seamless Interface between Work and the Rest of Life?', paper presented at the 22nd Nordic Sociology Congress, Malmö, Sweden.

DeFrank, R.S., Konopaske, R. and Ivancevich, J.M. (2000) 'Executive Travel Stress: Perils of the Road Warrior', *Academy of Management Executive*, 14: 58–71.

Dimberg, L.A., Striker, J., Nordanlycke-Yoo, C., Nady, L., Mundt, K.A. and Sulsky, S.I. (2002) 'Mental Health Insurance Claims among Spouses of Frequent Business Travellers', *Occupational and Environmental Medicine*, 59: 175–81.

Doyle, J. and Nathan, M. (2001) *Wherever Next? Work in a Mobile World*. London: Industrial Society.

Duncan, S., Edwards, R., Reynolds, T. and Alldred, P. (2003) 'Motherhood, Paid Work and Partnering: Values and Theories', *Work, Employment and Society*, 17: 309–30.

Elvin-Nowak, Y. (1998) *Flexibilitetens baksida: Om balans, kontroll och skuld i yrkesarbetande mödrars vardagsliv*. Stockholm: Psykologiska institutionen, Stockholms universitet.

Eriksson, B. (1998) *Arbetet i människors liv*. Monograph No. 66. Göteborg: Department of Sociology, Göteborg University.

Espino, C.M., Sundstrom, S.M., Frick, H.L., Jacobs, M. and Peters, M. (2002) 'International Business Travel: Impact on Families and Travellers', *Occupational and Environmental Medicine*, 59: 309–22.

Fisher, C. and Stoneman, B. (1998) 'Business on the Road', *American Demographics*, 20: 44–8.

Glass, J. and Estes, S. (1997) 'The Family Responsive Workplace', *Annual Review of Sociology*, 23: 289–313.

Grönlund, A. (2004) *Flexibilitetens gränser*. Umeå: Boréa.

Guest, D. (2002) 'Perspectives on the Study of Work–life Balance', *Social Science Information*, 41: 255–79.

Gustafson, P. (2005) *Resor i arbetet: En kartläggning av svenskarnas tjänsteresor 1995–2001*. Göteborg: Department of Sociology, Göteborg University.

Gustafson, P. (2006) 'Work-related Travel, Gender and Family Obligations', *Work, Employment and Society*, 20: 513–30.

Hochschild, A.R. (1997) *The Time Bind: When Work Becomes Home and Home Becomes Work*. New York: Metropolitan Books.

Hogarth, T., Hasluck, C., Pierre, G., Winterbotham, M. and Vivian, D. (2000) *Work-Life Balance 2000: Baseline Study of Work–Life Practices in Great Britain*. London: DfEE. www.dfes.gov.uk/research/data/uploadfiles/RR249.PDF.

Hyman, J., Scholarios, D. and Baldry, C. (2005) '"Daddy, I Don't Like These Shifts You're Working Because I Never See You": Coping Strategies for Home and Work', in D.M. Houston (ed.) *Work-Life Balance in the 21st Century*. Basingstoke: Palgrave Macmillan.

Ivancevic, J.M., Konopaske, R. and DeFrank, R.S. (2003) 'Business Travel Stress: A Model, Propositions and Managerial Implications', *Work and Stress*, 17: 138–57.

Jónasdóttír, A. (1994) *Why Women Are Oppres*sed. Philadelphia, PA: Temple University Press.

Kaufmann, V. (2002) *Re-thinking Mobility: Contemporary Sociology*. Aldershot: Ashgate.

Larsen, J., Urry, J. and Axhausen, K. (2006) *Mobilities, Networks, Geographies*. Aldershot: Ashgate.

Lassen, C. (2006) 'Aeromobility and Work', *Environment and Planning A*, 38: 301–12.

Leed, E.J. (1991) *The Mind of the Traveler: From Gilgamesh to Global Tourism*. New York: Basic Books.

Lewis, S. and Lewis, J. (1996) *The Work–Family Challenge*. London: Sage.

Markham, W.T., Bonjean, C.M. and Corder, J. (1986) 'Gender, Out-of-Town Travel, and Occupational Advancement', *Sociology and Social Research*, 70: 156–60.

Nolan, J. (2002) 'The Intensification of Everyday Life', in B. Burchell, D. Ladipo and F. Wilkinson (eds) *Job Insecurity and Work Intensification*. London: Routledge.

Peiperl, M. and Jones, B. (2001) 'Workaholics and Overworkers: Productivity or Pathology?', *Group and Organization Management*, 26: 369–93.

Presser, H.B. and Hermsen, J.M. (1996) 'Gender Differences in the Determinants of Work-Related Overnight Travel among Employed Americans', *Work and Occupations*, 23: 87–115.

Reeves, R. (2001) *Happy Mondays: Putting the Pleasure Back into Work*. Harlow: Momentum.

Reich, R. (2000) *The Future of Success: Work and Life in the New Economy*. London: Vintage.

Rutherford, S. (2001) '"Are You Going Home Already?" The Long Hours Culture, Women Managers and Patriarchal Closure', *Time and Society*, 10: 259–76.

Sheller, M. and Urry, J. (2006) 'The New Mobilities Paradigm', *Environment and Planning A*, 38: 207–26.

Smithson, J. and Stokoe, H. (2005) 'Discourses of Work–Life Balance: Negotiating Genderblind Terms in Organizations', *Gender, Work and Organization*, 12: 147–68.

Striker, J., Dimberg, L. and Liese, B.H. (2000) 'Stress and Business Travel: Individual, Managerial, and Corporate Concerns', *Journal of Organizational Excellence*, 20: 3–9.

Swarbrooke, J. and Horner, S. (2001) *Business Travel and Tourism*. Oxford: Butterworth-Heinemann.

Urry, J. (2002) 'Mobility and Proximity', *Sociology*, 36: 255–74.

Westwood, S., Pritchard, A. and Morgan, N.J. (2000) 'Gender-Blind Marketing: Businesswomen's Perceptions of Airline Services', *Tourism Management*, 21: 353–62.

Wolff, J. (1993) 'On the Road Again: Metaphors of Travel in Cultural Criticism', *Cultural Studies*, 7: 224–39.

15 Do mobile technologies enable work–life balance?

Dual perspectives on BlackBerry usage for supplemental work

Catherine A. Middleton

Introduction

This chapter explores the usage of mobile communication devices to support supplemental work. The 'anytime, anywhere' functionality of the devices makes them enormously convenient, and is thought to enhance people's work productivity, while facilitating work–life balance. But their always-on nature can lead to conflict when family members or others outside the users' work environment feel that work is spilling over into the users' non-work life.

With the help of texts from newspapers and magazines, the chapter investigates usage of a popular mobile device, the BlackBerry®, from the perspectives of users' families and friends, and of the users themselves. The contradictory interpretations are striking. Indeed, the very acts that define balance for BlackBerry users are clear signals of imbalance to those around them, resulting in strong opposition to the devices among non-users. Described as BlackBerry orphans (Rosman, 2006) and widows (Sokol, 2006; von Hahn, 2004), family members and friends express 'chagrin,' 'aggravation' 'disapproval' and 'ire' about the use of the device in their homes (and elsewhere).

The chapter shows how the behaviours that users adopt to improve their work–life balance result in the materialization of work, and taunt those in the non-work environment with 'absent presence.' As the uptake of 'mobile work-extending technologies' like BlackBerries is expected to rise in the future, the chapter outlines questions that should be addressed to help reduce the potential for work–life conflict.

Work, mobile technologies and work–life balance

There is a vast literature on telecommuting and telework, which provides the foundation for more recent studies on mobile work. 'Telecommuting' refers to a specific arrangement to work at home, reducing or eliminating the need to travel (commute) to work (Nilles, 1976). 'Telework' is used to describe 'remote work [that] involves the use of information and communication technologies' (Sullivan, 2003, p. 159). Many researchers consider the terms telework and telecommuting synonymously (Ellison, 1999). What is important in this context is that an explicit arrangement (voluntary or involuntary) is made between an employee and an employer that relocates some or all of his or her tasks to the home, from an office location (Felstead *et al.*, 2002; Fleetwood, 2007). These arrangements represent a substitution in the work environment, where employees give up some time in their offices and replace it with time spent working at home (Kraut, 1989). But the mobile work behaviours described here are not generally part of a formal, intentional relocation of work from one environment to another. Employees are not giving up their office space, instead they are extending their work environments to include spaces beyond the office. This is an important distinction (Kossek *et al.*, 2006), yet the supplemental nature of such work practices is not always reflected in studies on location of work (e.g. Felstead *et al.*, 2005; Hill *et al.*, 2003).

Bailyn (1988) describes this extension of work into home as 'overflow,' and notes that people have been bringing work home from the office for many, many years. New technologies allow knowledge workers to access, edit and create files, communicate with colleagues or clients, search for information and conduct other tasks from many locations outside their offices. Brown and O'Hara (2003, p. 1575) observe that mobile work 'makes place,' rather than 'taking place,' suggesting that any location can be made into a workplace by virtue of the fact that someone chooses to work there. The portability of work, and of technologies, allows employees to carry out 'supplemental work at home' (Venkatesh and Vitalari, 1992) but also extends the potential workplace to anywhere within the reach of mobile technology. In the past decade, supplemental work *at home* has given way to supplemental work *anywhere*.

The practice of working anywhere could easily be described as mobile work. Hislop and Axtell (2007) point out that mobility is not considered in the existing telework literature, but argue that mobile telework is becoming 'an increasingly important form of work' (p. 35). Mobile teleworkers move between home, office and 'locations beyond home and office' (p. 46), which include client premises and places visited for business travel. However, Hislop and Axtell do not appear to identify these spaces as locations for supplemental work. Other studies of mobile work (e. g. Brodt and Verburg, 2007; Brown and O'Hara, 2003) also exclude explicit discussion of mobile work conducted outside usual working hours.

Thus, while there are existing literatures on supplemental work at home, and on mobile work, it appears that limited academic attention has been paid to date to the phenomenon of mobile technologies being adopted in ways that allow supplemental work to move beyond the boundaries of home. One exception is Duxbury *et al.*'s (2005) research on 'work extension.' Their definition of work extension recognizes that much work is now done outside office hours (anytime) and at multiple locations outside the office (anywhere). Thus, extended work is supplemental work, but the definition no longer limits the location of supplemental work to the home. Personal digital assistants (PDAs), laptop computers, mobile email devices (e.g. BlackBerries) and home PCs are all considered work-extending technologies, and the technologies are becoming more prevalent among managerial and professional workers (Towers *et al.*, 2006).

As more people adopt extended work patterns, work is imposed on spaces and at times that were previously 'work free,' thus increasing the potential for role conflict. Conflict between work and non-work environments is not new (Greenhaus and Beutell, 1985; Lewis *et al.*, 2007) and it is addressed by an extensive literature (see Edwards & Rothbard, 2000, for a review of key concepts). However, much previous work on 'work–life' or 'work–family' balance in a telework environment (e.g. Golden *et al.*, 2006; Hill *et al.*, 2003; Madsen, 2003; Shumate and Fulk, 2004) does not reflect the pervasiveness or ubiquity of mobile technologies, nor does it fully reflect the supplemental nature of work that is extending beyond office hours and office boundaries. When supplemental and mobile work converge to create an anytime, anywhere, always-on work environment, the potential for conflict and imbalance is exacerbated (Menzies, 2005).

Balance means different things to different people, and the distinction between 'work' and 'life' is problematic. The description of 'family' as being the core of life outside work is too narrow (Ransome, 2007), while focusing on family alone as the key component to life outside work excludes leisure and other non-family, non-work responsibilities (e.g. contribution to local communities) (Guest, 2002). For expediency, however, in this chapter participants in the non-work sphere of individuals' lives are referred to as 'friends' and 'family,' and the non-work sphere is simply referred to as 'life.'

Boundaries delineate spaces in people's lives (e.g. work, home, 'third places') (Nippert-Eng, 1996), and individuals assume various roles (e.g. parent, partner, employee) within these bounded spaces (Ashforth *et al.*, 2000). Few people are able to completely segment the spaces in their lives, thus there is potential for 'role conflict' as the demands of one role compete with the demands of another (Kahn *et al.*, 1964). Conflict between work and family is a particular concern (Duxbury *et al.*, 1994, 2003; Jacobs and Green, 1998).

Clark's (2000) 'work/family border theory' builds on the assumption that work and family occupy separate, yet related spheres. Each sphere has its own rules and culture, and there is a border between the two. The theory illustrates the integrated, interdependent nature of work and family life, and moves beyond simple observations of role conflict to frame the search for work–life balance as a dynamic, ongoing set of negotiations that occur whenever individuals cross the border between their work and life domains. Clark outlines various propositions that explain how borders can facilitate or inhibit work–life balance. Assessment of work–life balance is influenced by a person's role in a domain, and by the strength of the border between domains. She suggests that individuals who are central participants in a domain (i.e. strongly identify with the domain and influence its environment) have more control over the border than peripheral participants. As such, the expectations of the life domain may be set by a worker's family, and not be in line with the expectations of the work domain-centric border-crosser (and vice versa). However, it is also proposed that 'other domain awareness' influences work-family balance, so that when domain members are aware of the demands of the other domain, they are more supportive of the border-crosser.

Clark's theory does not consider the impact of work-extending technologies on work–life balance, but it is instructive to think of these technologies crossing borders with their users. Work-extending technologies are thought to enable improved productivity (Smith, 2005) and help workers to achieve greater balance between work and non-work environments (Cousins and Robey, 2005; Schlosser, 2002), but others suggest that this technology-enabled extension of work is invasive and counterproductive (Hallowell, 2005; Jackson, 2007). What is of interest in this chapter is the adoption of work-extending technologies, like the BlackBerry, that are portable, mobile and work almost everywhere, regardless of users' work–life borders.

As early as 2000, the term 'Crackberry' was being used on Wall Street to describe the addictive nature of these devices (Haines, 2000). While the question of whether BlackBerry aficionados are clinically addicted to their devices remains open (see Porter and Kakabadse, 2006, for a discussion of technology addiction in the workplace), there is no doubt that BlackBerries are deeply embedded in the daily lives of many of their users, and can trigger conflicts about work boundaries (Middleton, 2007; Middleton and Cukier, 2006). This chapter explores the relationship between BlackBerry adoption and perceptions of work–life balance. Unlike Middleton and Cukier's previous work, this study encompasses the dual perspectives of BlackBerry users and their families and friends, recognizing the importance of assessing balance and work–life conflict from both sides (Hill *et al.*, 1996). It builds on earlier work by Schlosser (2002), Mazmanian *et al.* (2005), Towers *et al.* (2006), and Golden and Geisler (2007) to further explore work–life balance challenges faced by knowledge workers when employing mobile devices that offer users ubiquitous connectivity to the office and support supplemental work at any time and from any location.

In North America, the BlackBerry has become the device of choice for mobile email. First attracting public notice for providing communication in New York City on September 11, 2001 after much of the telecommunications infrastructure failed (see, for example, 'Downtown BlackBerry E-Mail Repository, 2007'), the BlackBerry experienced slow but steady growth in subscriptions for its first few years. By early 2004, there were more than one million BlackBerry subscribers, and by mid-2005, three million people had subscribed (Research in Motion, 2004; Research in Motion, 2005) to this 'iconic pocket-sized e-mail device' (Economist Staff, 2005). A patent dispute in 2006 that threatened to shut down BlackBerry service caused much consternation among users as they faced the potential loss of their devices (Parks, 2006; Smith, 2006). Although rare, disruptions in service are headline news (e.g. Vascellaro *et al.*, 2007). As of late 2007, there were more than 10.5 million subscribers (Research in Motion, 2007), with growth estimated at one million subscribers every three months (Sørensen, 2007).

The BlackBerry's reputation, and continued success, rests upon its highly reliable, secure and user-friendly email service – 'It's small and it works' (Estates Gazette Staff, 2005). The device is a PDA and a mobile phone, and provides 'push' email functionality, delivering messages as they are received without the need for users to take action to connect to the internet. In many countries, before even stepping off an airplane, travelers can send and receive email effortlessly by just turning on their BlackBerries. This simple device has become indispensable for legions of businesspeople around the world. It allows people to check their email anywhere, and to respond to messages in an unobtrusive manner. It also makes it very easy for individuals to carry their work with them, and to engage in work activities in locations and at times that were previously 'off limits.'

The following section presents data showing how BlackBerries can serve to support supplemental work, explaining how the devices help users balance their work–life responsibilities. It also shows how this assessment of work–life balance is not shared by BlackBerry users' friends and family, who perceive the devices to be disruptive, distracting and over-used.

BlackBerry usage data

The data that follow are drawn from popular press accounts of BlackBerry usage in the past two years (2005–2007), as catalogued in the *Factiva* database. After a search on the term 'BlackBerry' yielded almost 60,000 'hits,' the more restrictive term 'Crackberry' was used to limit the search. While this approach excluded relevant articles about BlackBerry usage that did not mention the word Crackberry, it does furnish a good sample of articles that address the tensions created as mobile technologies enable work to spill over into other aspects of people's lives.

From a starting point of more than 1,000 articles, a research team removed duplicates and irrelevant articles, resulting in a final compilation of just over 200 articles that discussed various aspects of BlackBerry (and other mobile device) usage in individuals' daily lives. The team then indexed the articles[1] in a bibliographic software program and exported the texts into a qualitative data analysis program for thematic analysis employing a semi-structured coding protocol.

It might be argued that BlackBerry usage behaviours deemed newsworthy are extreme ones, and not representative of 'ordinary' BlackBerry users going about their daily lives. But the vivid examples presented here do show the conflicts inherent in adopting mobile technologies to extend supplemental work practices, and provide a focal point for discussing the implications of the continued uptake of work-extending technologies. While the results may not be generalizable, the anecdotes presented here are consistent with descriptions of BlackBerry usage in a small-scale study of Canadian BlackBerry users conducted in 2005 (Middleton and Cukier, 2006; Middleton *et al.*, 2005), and afford insights into users' and non-users' experiences of 'mobile work-extending technologies.'

In the section below, data are presented to show how BlackBerries are deployed for supplemental work away from the office. Descriptions of how the devices enable work–life balance for the users are supplied, followed by evidence from non-users that offers a contrary perspective on the device's role in balancing the work and non-work spheres.

Location of use

BlackBerry users are described as 'the ones hunched over like squirrels with a walnut, thumbs flying manically, even at weddings, funerals and the movies.' A 'devoted' user reported accessing his BlackBerry during his wife's stepfather's funeral, a Congressman was observed spending 'a great deal of time on his BlackBerry during [Ash Wednesday] service and prayer, both reading emails and sending emails.'

Some take their BlackBerries into the shower ('keep[ing] it within view but dry'), and there are reports of people who 'accidentally dropped the device in the toilet.' One man described how he'd 'fallen asleep with it in [his] hands, read it as [he] ate, watched TV, waited in line, and while playing soccer with [his] son.' Others spend time at their children's concerts, baseball games or swim practices with BlackBerries in hand, pleased that they can 'still be at work!'

BlackBerries are frequently found at the meal table. It seems that no special occasion is exempt, as people confess to 'us(ing) it at Passover dinner,' and 'interrupting the turkey dinner, mince pies and festivities on Christmas day to check their BlackBerry for email messages and keep tabs on the company's IT operations.' One woman 'caught her husband e-mailing under the table during her Valentine's Day dinner,' while another found her companion checking email throughout their first date. These practices are captured nicely by cartoonist Philip Street in his *Fisher* strip.[2] See Figure 15.1.

In describing the factors that led up to his divorce, a man says 'the thing that really brought it home to me was we were in an intimate moment in bed, and I lifted up my head and I caught my wife checking her e-mail on the BlackBerry.' Not an isolated incident, a doctor reported being asked by a patient 'whether [he] thought it was abnormal that her husband brings the BlackBerry to bed and lays it next to them while they make love.'

A woman describes a dream 'about squirrels eating acorns.... And then I woke up, and it was my husband, the tap, tap, tap, tap on the BlackBerry.' A man reports that BlackBerry is 'the last thing I check before going to sleep and the first thing I touch in the morning.' Some people even access it in the middle of the night, including one man who regularly checked email while getting up in the night with his newborn daughter.

Figure 15.1 © Philip Street in his *Fisher* strip (used with permission).

There are many reports of drivers engaged with BlackBerries ('It is actually scary to see people driving in their cars receiving and sending e-mails'), and the devices also accompany their users on vacation. BlackBerries can be found on the golf course, poolside or at the beach. One man took his BlackBerry to Maui for his tenth anniversary celebration, and another 'went to Disneyland last year accompanied by his wife, their two children and his BlackBerry. According to his wife, the BlackBerry drained much of the magic from the Magic Kingdom.'

User perceptions of work–life balance

BlackBerries furnish workers with a '24/7' connection to their offices, and there is a strong sentiment that the devices help to achieve balance in users' lives. 'I like to be connected,' says a small business owner.

> I don't know what I would do without it. And I'm much more likely to take vacation because of it. I have more work/life balance because I carry my Treo [a Palm Pilot product with similar functionality to the BlackBerry]; I feel less need to be in the office.

A lawyer describes how his BlackBerry allows him to 'go places and do things and still stay on top of my work … keep[ing] tabs on the office, while hanging out with his kids.'

BlackBerries allow their users to be efficient, while spending time with friends and family – 'If we're standing in line for 40 minutes waiting for a ride [at Disneyland], I don't see why I can't answer my e-mail,' says one user. When his son made the Little League all-star team, a man enthused that 'the BlackBerry allowed me to go to the game and still deal with some real-time issues we had in the office.' A 2006 survey by recruitment firm Korn/Ferry found that

> More than one-third of 2,300 executives surveyed in 75 countries believed they spent too much time connected to communications devices. But more than three-quarters, or 77 percent of respondents, said they believe mobile communication devices primarily enhance their work/life balance rather than impede it.

An alternative perspective on work–life balance

Many people, especially friends and family of BlackBerry users, do not share the belief that the devices create balance. This quote expresses a common sentiment – 'She hates that he's a BlackBerry fiend, especially when he argues that using it leaves more time for family.' The important people in users' lives are not shy in expressing their opinions about BlackBerry use in their environments. While a four-year old expressed her displeasure at her mother's BlackBerry usage by simply hiding the device, her seven-year-old brother was more sophisticated in trying to flush it down a toilet. Immersing BlackBerries in water

Figure 15.2 Alex cartoon by Peattie and Taylor. *Alex* appears in the *Daily Telegraph*.

seems to have broad appeal. 'The winner of the British version of *The Apprentice*, a reality TV show, has admitted that his wife has threatened to flush his BlackBerry down the toilet,' a threat repeated by other users' spouses. One wife 'wanted to pick it up and throw it into the swimming pool' (while on vacation) while another 'tried to throw it off the boat when [they] were on [their] honeymoon.' The *Alex* comic strip[3] regularly captures the frustrations of BlackBerry lovers' families, as seen in Figure 15.2 above, (www.alexcartoon.com/index.cfm?cartoons_id=2238, 22 May 2004).

Throwing the BlackBerry out a window was also suggested by an irate wife who felt ignored by her husband. A husband remarked that he would not use his BlackBerry at Christmas, for fear of watching his 'BlackBerry crackling away on the fire along with the Yule log.'

In some households, family members have adopted 'rules of engagement' around BlackBerries. This may mean a ban on the device on weekends, or in restaurants and the bedroom. Children help to discourage their parents' BlackBerry usage, 'begging' them to stop using it at the table. One woman was surprised when her daughter 'literally applauded her decision to leave her BlackBerry behind when vacationing.' Nevertheless, some people continue to access their BlackBerries, even when it is very clear that such usage is not acceptable to others. Fearing discovery, they hide their devices from spouses or family members but insist that their behaviours are justified. One man explains that 'his BlackBerry actually alleviates marital tension by allowing him to secretly check his email and get work done during vacations with his wife.' Another individual reports that checking his BlackBerry on vacation (while hiding in the bathroom to do so) resulted in 'A relaxed me, an unsuspecting girlfriend, a holiday success.'

Analysis

The data presented here show the pervasive usage of BlackBerries, and demonstrate the conflicting assessment of the value of such devices. BlackBerries do enable people to be connected to their work from anywhere, at any time. This connectivity provides users great comfort because it allows them to remain in contact with their jobs while attending to other aspects of their lives. While there is no doubt that many people feel pressured to remain connected to work at all hours, with some organizational cultures reinforcing and validating this expectation (Middleton, 2007), users are adamant that their BlackBerries lend them freedom, and contribute to a better work–life balance by allowing them to spend more time with friends and family. But their friends and family often resent the presence of the BlackBerry, seeing it as a means for workers to extend their work into spaces where work is not welcome. Rather than interpreting this as work–life balance, friends and family view anytime, anywhere BlackBerry usage as always-on work. Rather than experiencing less conflict as a result of being able to better manage their work and life commitments, BlackBerry users may face increased conflict, as their friends and family actively resist the device. BlackBerries have been successful because they can turn any place into a workplace, which is exactly the reason why they are reviled by those who want to contain work within well-defined, agreed upon boundaries.

Clark's (2000) work/family border theory (described earlier) offers some insights to help understand the data presented above. Of interest in this chapter is the border-crossing from the work domain into the family (non-work) domain, where work–family spillover is possible. The data show that, when BlackBerry users cross from the work to the life (non-work) sphere, they frequently bring their BlackBerries 'over the border.' They are met by the border-keeper, generally a spouse or significant other, as well as other domain members (e.g. children). It is expected that upon crossing the border (which may be physical, temporal or psychological), 'domain-relevant behavior' (Clark, 2000, p. 756) takes place.

Applying the concept of border-crossing to the data presented above generates insights related to two themes. The first theme is described as the materialization of work, in which a specific artifact, the BlackBerry, permeates the work–life border to bring work into what is understood to be a non-work environment. The second theme relates to the idea of 'absent presence' (Gergen, 2002), and can be seen here as a form of taunting. Given its visibility and popularity, the BlackBerry has garnered more attention than other devices, and it is likely a harbinger for more widespread uptake of mobile work-extending technologies. It is suggested that the observations made here are not device-dependent, but apply wherever mobile technologies are adopted to facilitate anytime, anywhere supplementary work.

Materialization of work

Border theory suggests that there are acceptable behaviours for each sphere, and that, when a person crosses the border, he or she transitions to the norms of the sphere just entered. Ashforth *et al.* (2000) note that these crossings involve exiting one role and taking up another. The adoption of mobile technologies reduces the likelihood that such role exit will actually occur when moving across the work–life border, as the demands of the work role can continue to be met by utilizing mobile technologies in the life sphere. As such, a BlackBerry can be understood as a very visible manifestation of work and of permeable work–life borders. When the device is taken across the work–life border, it affords a clear indication that the user remains linked to the work domain even though he or she is physically present in the non-work domain. Even if the device is turned off, its mere presence signals that work is possible. Users argue that this provides them with the flexibility to attend to their non-work lives without neglecting work duties, but from the perspective of the domain members, this materialization of work shows that users have not left the work domain.

Prior to the widespread adoption of mobile devices, it was easier to contain work within physical and temporal boundaries. While spillover of work into the non-work domain has always been a potential source of conflict, what has changed with the uptake of mobile work-extending technologies is that temporal and physical boundaries are more easily breached. It is easy to take a BlackBerry to a social event (dinner party, baseball game) or to check email while lying in bed or while sitting by the pool on vacation. Users view such behaviours as freeing themselves from the physical constraints of the office, but for their friends and family, work is now visibly occupying times and spaces in the non-work domain that were previously off-limits. The device that enables this extension of work acts as a 'lightning rod,' attracting attention to the presence of work. Despite users' best efforts to be discreet when accessing BlackBerries in 'inappropriate' settings, their presence draws attention to work. Because it is so pervasive, and acts as a persistent visual reminder that work has infiltrated the non-work domain, the BlackBerry has become an obvious target for criticism and a flashpoint for work–family conflict. The device may well act as a proxy for broader dissent about differential expectations regarding work–life balance, increasing the intensity of resistance to the device and explaining why its very appearance can provoke such ire and emotion from users' friends and families.

Absent presence: how mobile devices taunt non-users

Not only does the BlackBerry bring a visible manifestation of work into the home and other non-work environments, it can also psychologically remove users from the non-work environment and return them to a work mindset. As has been mentioned, BlackBerry users feel that the device allows them to balance work and life domains, because they can attend to work needs while outside the workplace. But although physically present in the non-work domain, whenever people engage with their BlackBerries, they are removing themselves from their present environment and focusing their attention elsewhere. Described by Gergen (2002) as 'absent presence' and by Fortunati (2002) as 'present absence' this behaviour taunts those around the user by giving the appearance of attention to, or participation in the non-work domain, while the user actually remains grounded in the work domain.

Users pride themselves on the fact that their BlackBerries allow them to attend events and participate in activities that they would have missed in the days before mobile technologies, yet arguably, they are still missing such events by engaging with their devices, rather than with their physical environment. In the past, people with heavy work commitments would have met these commitments by staying at the office to complete the work, or by confining their work to a specific location within their non-work domain (e.g. a home office), and not participating in the non-work domain. BlackBerries allow the work to be done anywhere, satisfying users that they are achieving balance, but frustrating their friends and family by making it more obvious that work is spilling over into non-work times and spaces. Given the particular reactions that BlackBerry use in the non-work domain provokes, it is understandable that non-users might interpret this behaviour as taunting. In the name of participating in activities with families and friends, BlackBerry users join the non-work environment, and promote the appearance of being engaged with it, but can at any time 'step out' of the environment to return to work. From the perspective of BlackBerry users, the guilt of missing an activity is removed or at least mitigated, but from the perspective of family and friends, it appears that the BlackBerry exacerbates the awareness of work–life imbalance.

Discussion and conclusions

The anecdotes of BlackBerry usage presented here show how actions that knowledge workers take to balance their work activities with their personal lives can result in conflict. By materializing work, mobile work-extending technologies like BlackBerries can become the centre of attention when employed outside the office, and provide a focal point for discontent among friends and family members. Likewise, efforts at being present in the non-work environment are not always met with approval. Although the workers make a special effort to engage with their friends and family by participating in events and activities, the fact that they bring their BlackBerries with them triggers resentment. Rather than appreciating the worker's presence in the non-work environment, attention is focused on the absences created when the worker engages with his or her job through a mobile device.

It is likely that the workers do not fully understand their friends' and family members' disdain for their devices (and equally likely that friends and family do not understand the demanding nature of the work environment that does expect workers to be connected and available outside business hours). Towers *et al.* (2006) found that heavy users of work-extending technologies believed that their families understood their need to work during family time and, although they recognized that heavy usage could be problematic, individuals felt that they were doing a good job of controlling the extent to which their technology use was spilling over into their personal lives. This justification of individual work practices indicates that workers believe that their approach of combining work and non-work activities is both effective and appropriate.

This approach to work–life balance is comparable to the 'integrating the self' repertoire identified by Golden and Geisler (2007), in which PDA users explicitly utilized their devices to transcend, rather than contain, work–life boundaries. Felstead and Jewson (2000) identify segregated and integrated approaches to creating work–life boundaries. The integrated approach, which was adopted by the BlackBerry users described here, is based on weak temporal and spatial separation of work and non-work domains. In their study comparing different types of mobile work, Hislop and Axtell (2007) showed that an integrated approach delivered less work–life balance than a segregated approach.

This study provides no point of comparison to determine whether a more segregated approach to BlackBerry adoption would have resulted in less work–life conflict, but it does show that the integrated approach that was adopted did not sit well with friends and family. This is an interesting finding, because one of the key affordances of mobile work-extending technologies like BlackBerries is that they allow people to integrate their home and work lives, and to maintain open boundaries between the two. This study suggests that, while this works for the BlackBerry users, it may not work for those around them. It is possible that the covert uses are a response to the shortcomings of an integrated approach, allowing individuals to avoid disapproval and conflict by reverting to absence and secrecy when conducting their work in non-work domains.

BlackBerries and other mobile work-extending technologies are still relatively new, and it is likely that the ways in which they are deployed will evolve over time. There is some evidence of users adopting more structured approaches to keep their work and personal lives in balance (Jackson, 2007), but the patterns portrayed here are the dominant ones at present. As noted earlier, for many, the appeal of the BlackBerry or other mobile devices is that they do enable anytime, anywhere work, functionality which has been constructed by users as a means of controlling their busy, demanding lives and enhancing work–life balance. As such, it is expected that the usage patterns documented here and the conflict such usage engenders will continue. This raises a number of questions to be considered by those adopting mobile technologies to support supplemental work, and by researchers interested in the intersection of mobility and supplemental work.

- What are the longer-term implications of work–life conflict that is exacerbated by the adoption of mobile devices? Are there ways of mitigating the conflict? What actions could be taken to achieve better fit between the employees' real needs to remain connected to work while away from the office, and the demands of their non-work environments? Can people learn to temper their addict-like attention to their devices, while those around them accept that some usage is necessary? Are there alternatives to covert use that meet the needs of users and their friends and families?
- What are the broader forces driving this compulsive attachment to mobile work-extending technologies? Are the devices truly addictive, or do users exhibit signs of being addicted to their work? What can be learned from an extensive reading of the literature on workaholism (see, for example, Burke,

2006; Kofodimos, 1993; Porter, 2006)? For example, do choices that users make with respect to favouring their work domains over non-work ones suggest deeper issues regarding their relationships with each domain?

- What are the broader cultural and societal forces driving such behaviours? Why do organizations support uses that can have negative impacts on their employees' personal lives (and potentially reduce overall productivity and effectiveness)? Why do employees feel such compulsion to remain connected to their offices and to work all the time? To what extent is supplemental work really necessary?

This chapter contributes to our understanding of technology-enabled mobile work by affording insights into the usage of mobile technologies to support supplemental work. By definition, supplemental work occurs outside the office, and with the advent of ubiquitous, user-friendly communication devices, it can be, and is, done from anywhere, at any time. The chapter indicates that claims that mobile technologies facilitate work–life balance are one-sided, and applies border theory to explain how current uses can increase work–life conflict by materializing work and taunting family and friends with absent presence. Given that the adoption of mobile work-extending technologies is expected to increase, it is important that all those affected consider how to make such usage more favourable to all.

There are more questions than answers at present. The convergence of supplemental work and mobile technologies raises complex issues that require much more nuanced analysis and a greater grounding in the literature than can be furnished within a single book chapter. Issues of gender and power were not addressed here but must be considered. It is also important to determine the extent to which individuals and organizations are willing to move toward an environment of always-on, anytime, anywhere work. What do people really want, and how can they ensure that their needs are not subsumed by corporate agendas and unfettered, uncritical adoption of technologies? In 1988, Bailyn wrote that 'Information technology makes it possible to free work from the constraints of location and time' (p. 149). Today the challenge is to free location and time from the constraints of work.

Notes

1 Specific references are not provided for the data presented here. The author can provide full references for quotations upon request. Identifying information on individual BlackBerry users has been removed from the data.

2 See www.philipstreet.com/fisher/archives.html for archives of the *Fisher* strip. BlackBerries feature in the comic strip 7, 8 and 9 April 2005, and again 20 March through 23 March 2007.

3 A search for 'BlackBerry' in the *Alex* archive www.alexcartoon.com/index.cfm?section=archive_search returns many insightful comics about BlackBerry usage.

References

Ashforth, B.E., Kreiner, G.E. and Fugate, M. (2000) 'All in a Day's Work,' *Academy of Management Review*, 25: 472–91.

Bailyn, L. (1988) 'Freeing Work from the Constraints of Location and Time,' *New Technology, Work and Employment*, 3: 143–52.

Brodt, T.L. and Verburg, R.M. (2007) 'Managing Mobile Work – Insights from European Practice,' *New Technology, Work and Employment*, 22: 52–65.

Brown, B. and O'Hara, K. (2003) 'Place as a Practical Concern of Mobile Workers,' *Environment and Planning A*, 35: 1565–87.

Burke, R.J. (2006) *Research Companion to Working Time and Work Addiction*. Cheltenham: Edward Elgar.

Clark, S.C. (2000) 'Work/Family Border Theory: A New Theory of Work/Family Balance,' *Human Relations*, 53: 747–70.

Cousins, K.C. and Robey, D. (2005) 'Human Agency in a Wireless World: Patterns of Technology Use in Nomadic Computing Environments,' *Information and Organization*, 15: 151–80.

Downtown BlackBerry E-Mail Repository (2007) Retrieved 28 July, from www.911digitalarchive.org/email/downtown_blackberry.html.

Duxbury, L., Higgins, C. and Coghill, D. (2003) *Voices of Canadians: Seeking Work–Life Balance*. Hull, QC: Human Resources Development Canada.

Duxbury, L., Higgins, C. and Lee, C. (1994) 'Work-Family Conflict: A Comparison by Gender, Family Type, and Perceived Control,' *Journal of Family Issues*, 15: 449–66.

Duxbury, L., Thomas, J.A., Towers, I. and Higgins, C. (2005) 'From 9 to 5 to 24 and 7: How Technology Has Redefined the Workday,' report prepared for Industry Canada, Ottawa.

Economist Staff (2005) 'The CrackBerry Backlash,' *The Economist*, 375: 78.

Edwards, J.R. and Rothbard, N.P. (2000) 'Mechanisms Linking Work and Family: Clarifying the Relationship between Work and Family Constructs,' *Academy of Management Review*, 25: 178–99.

Ellison, N.B. (1999) 'Social Impacts: New Perspectives on Telework,' *Social Science Computer Review*, 17: 338–56.

Estates Gazette Staff (2005) 'Season of Fruitful Technology,' *Estates Gazette*, 15 October: 89.

Felstead, A. and Jewson, N. (2000) *In Work, at Home: Towards an Understanding of Homeworking*. London: Routledge.

Felstead, A., Jewson, N. and Walters, S. (2005) 'The Shifting Locations of Work: New Statistical Evidence on the Spaces and Places of Employment,' *Work, Employment & Society*, 19: 415–31.

Felstead, A., Jewson, N., Phizacklea, A. and Walters, S. (2002) 'The Option to Work at Home: Another Privilege for the Favoured Few?,' *New Technology, Work and Employment*, 17: 204–23.

Fleetwood, S. (2007) 'Why Work–Life Balance Now?' *International Journal of Human Resource Management*, 18: 387–400.

Fortunati, L. (2002) 'The Mobile Phone: Towards New Categories and Social Relations,' *Information, Communication & Society*, 5: 513–28.

Gergen, K.J. (2002) 'The Challenge of Absent Presence,' in J.E. Katz and M.A. Aakhus (eds) *Perpetual Contact: Mobile Communication, Private Talk, Public Performance*. Cambridge, MA: Cambridge University Press, pp. 227–41.

Golden, A.G. and Geisler, C. (2007) 'Work–Life Boundary Management and the Personal Digital Assistant,' *Human Relations*, 60: 519–51.

Golden, T.D., Veiga, J.F. and Simsek, Z. (2006) 'Telecommuting's Differential Impact on Work-Family Conflict: Is There No Place Like Home?' *Journal of Applied Psychology*, 91: 1340–50.

Greenhaus, J.H. and Beutell, N.J. (1985) 'Sources of Conflict between Work and Family Roles,' *Academy of Management Review*, 10: 76.

Guest, D.E. (2002) 'Perspectives on the Study of Work–Life Balance,' *Social Science Information*, 41: 255–79.

Haines, M. (2000) Research in Motion – CFO – Interview: Transcript,' 29 September. CNBC/Dow Jones Business Video.

Hallowell, E. (2005) 'Overloaded Circuits: Why Smart People Underperform,' *Harvard Business Review*, 83: 54–62.

Hill, E.J., Ferris, M. and Martinson, V. (2003) 'Does It Matter Where You Work? A Comparison of How Three Work Venues (Traditional Office, Virtual Office, and Home Office) Influence Aspects of Work and Personal/Family Life,' *Journal of Vocational Behavior*, 63: 220–41.

Hill, E.J., Hawkins, A.J. and Miller, B.C. (1996) 'Work and Family in the Virtual Office: Perceived Influences of Mobile Telework,' *Family Relations*, 45: 293–301.

Hislop, D. and Axtell, C. (2007) 'The Neglect of Spatial Mobility in Contemporary Studies of Work: The Case of Telework,' *New Technology, Work and Employment*, 22: 34–51.

Jackson, M. (2007) 'Keeping Work Where It Belongs,' *Boston Globe*, 3 June: G19.

Jacobs, J.A. and Green, K. (1998) 'Who Are the Overworked Americans?' *Review of Social Economy*, 56: 442–59.

Kahn, R., Wolfe, D., Quinn, R., Snoek, J. and Rosenthal, R. (1964) *Organizational Stress: Studies in Role Conflict and Ambiguity*. New York: Wiley.

Kofodimos, J.R. (1993) *Balancing Act: How Managers Can Integrate Successful Careers and Fulfilling Personal Lives* 1st edition. San Francisco, CA: Jossey-Bass.

Kossek, E.E., Lautsch, B.A. and Eaton, S.C. (2006) 'Telecommuting, Control, and Boundary Management: Correlates of Policy Use and Practice, Job Control, and Work–Family Effectiveness,' *Journal of Vocational Behavior*, 68: 347–67.

Kraut, R.E. (1989) 'Telecommuting: The Trade-Offs of Home Work,' *Journal of Communication*, 39: 19–47.

Lewis, S., Gambles, R. and Rapoport, R. (2007) 'The Constraints of A "Work–Life Balance" Approach: An International Perspective,' *International Journal of Human Resource Management*, 18: 360–73.

Madsen, S.R. (2003) 'The Effects of Home-Based Teleworking on Work–Family Conflict,' *Human Resource Development Quarterly*, 14: 35–58.

Mazmanian, M.A., Orlikowski, W.J. and Yates, J. (2005) 'Crackberries: The Social Implications of Ubiquitous Wireless E-Mail Devices,' in C. Sørensen, Y. Yoo, K. Lyytinen and J.I. DeGross (eds) *Designing Ubiquitous Information Environments: Socio-technical Issues and Challenges*. New York: Springer, pp. 337–44.

Menzies, H. (2005) *No Time: Stress and the Crisis of Modern Life*. Vancouver, BC: Douglas & McIntyre.

Middleton, C.A. (2007) 'Understanding the BlackBerry as an Instrument of Organizational Culture,' *Continuum: Journal of Media and Cultural Studies*, 21: 165–78.

Middleton, C.A. and Cukier, W. (2006) 'Is Mobile Email Functional or Dysfunctional? Two Perspectives on Mobile Email Usage,' *European Journal of Information Systems*, 15: 252–60.

Middleton, C.A., Scheepers, H. and Cukier, W. (2005) 'Exploring the Contradictions of Mobility: A Case Study of BlackBerry Users in Canada,' in J.Y.L. Thong and K.Y. Tam (eds) *Proceedings of the Hong Kong Mobility Roundtable*. Hong Kong: HKUST, pp. 299–309.

Nilles, J.M. (1976) *The Telecommunications-Transportation Tradeoff: Options for Tomorrow*. New York: Wiley.

Nippert-Eng, C.E. (1996) *Home and Work: Negotiating Boundaries through Everyday Life*. Chicago, IL: University of Chicago Press.

Parks, B. (2006) 'Addiction, Withdrawal and the BlackBerry,' *Star-Ledger*, 29 January: 1.

Porter, G. (2006) 'Profiles of Workaholism among High-tech Managers,' *Career Development International*, 11: 440–62.

Porter, G. and Kakabadse, N.K. (2006) 'HRM Perspectives on Addiction to Technology and Work,' *Journal of Management Development*, 25: 535–60.

Ransome, P. (2007) 'Conceptualizing Boundaries between "Life" and "Work,"' *International Journal of Human Resource Management*, 18: 374–86.

Research in Motion (2004) *Annual Report*. Retrieved 14 February, 2005, from www.rim.com/investors/pdf/2004rim_ar.pdf.

Research in Motion (2005) *BlackBerry Subscribers Surge to over Three Million*. Retrieved 1 September, 2005, from www.blackberry.com/news/press/2005/pr-09_05_2005–01.shtml.

Research in Motion (2007) *Research in Motion Reports Second Quarter Results*. Retrieved 28 October, 2007, from www.rim.net/news/press/2007/pr-04_10_2007–01. shtml.

Rosman, K. (2006) 'BlackBerry Orphans: The Kids Fight Back; Heavy Use of E-Mail Gadgets Is Spawning a Generation of Resentful Children,' *Globe and Mail*, 9 December: B15.

Schlosser, F.K. (2002) 'So, How Do People Really Use Their Handheld Devices? An Interactive Study of Wireless Technology Use,' *Journal of Organizational Behavior*, 23: 401–23.

Shumate, M. and Fulk, J. (2004) 'Boundaries and Role Conflict When Work and Family Are Colocated: A Communication Network and Symbolic Interaction Approach,' *Human Relations*, 57: 55–74.

Smith, D. (2006) 'BlackBerry Addicts Fear Total Blackout,' *The Observer*, 5 February: 22.

Smith, I. (2005) 'Social-Mobile Applications,' *Computer*, 38: 84–5.

Sokol, J. (2006) 'The Tale of a Man and His CrackBerry, Er, BlackBerry,' *Orange County Register*, 5 July.

Sørensen, C. (2007) CEO Balsillie Shrugs Off 'BlackBerry Killer,' *Toronto Star*, 7 July: B1.

Sullivan, C. (2003) 'What's in a Name? Definitions and Conceptualisations of Teleworking and Homeworking,' *New Technology, Work and Employment*, 18: 158–65.

Towers, I., Duxbury, L., Higgins, C. and Thomas, J. (2006) 'Time Thieves and Space Invaders: Technology, Work and the Organization,' *Journal of Organizational Change Management*, 19: 593–618.

Vascellaro, J.E., Yuan, L., Sharma, A. and Rhoads, C. (2007) 'Disconnected and Discombobulated – "CrackBerry" Outage Forces Users to Go Cold Turkey; Pause in "Infernal Buzzing",' *The Wall Street Journal*, 19 April: B1.

Venkatesh, A. and Vitalari, N.P. (1992) 'An Emerging Distributed Work Arrangement: An Investigation of Computer-based Supplemental Work at Home,' *Management Science*, 38: 1687–706.

von Hahn, K. (2004) 'I'm a Blackberry Widow,' *Globe and Mail*, 1 May: L3.

Part V

Public policy

16 Mobile work and challenges for public policy

Dan Wheatley, Irene Hardill and Anne E. Green

Introduction

This chapter outlines and considers some of the key public policy implications related to mobile working. It draws on a range of evidence from the academic and policy literature, but particularly on recent empirical research undertaken in the UK (Wheatley, 2008).

The public policy implications considered include infrastructure development, regeneration and spatial planning, but a particular focus is placed on the work–life balance agenda. Potentially, mobile working has both positive and negative impacts on mobile workers, on their colleagues, on other household members, on their homes, on their workspaces and on places. Although mobile work is 'lived' by a wide range of workers, the focus here is on managers and professionals, many of whom are characterized by some degree of autonomy over when and where they undertake their work tasks (Hardill, 2002), and so among whom mobile working practices are especially prevalent. Typically, their jobs entail a high degree of commitment and are intrinsically demanding in character (Rapoport and Rapoport, 1976). Moreover, labour market restructuring, changes in employment relationships and trends in the nature and spatial configuration of managerial and professional work mean that the working environment and spatial mobility demands for many are different to those experienced in the past (see Table 16.1 for a summary of trends and developments).

This chapter is organized into four substantive sections, each considering the public policy challenges of mobile work from a different perspective: first, work-related travel; second, workspace/place; third, mobile work practices; and fourth, home–work dynamics. Reference is made to the key policy drivers affecting mobile working practices, including the labour market, work–life balance, planning, housing and transport policy. Here it is salient to note that mobile work is shaped by the action of policy makers at a number of spatial scales, from the local to supranational.

The focus of this chapter is largely on the UK situation, but where possible evidence from other countries is drawn upon.

Table 16.1 The new world of work for managers and professionals

Work organization	*Spatial reconfiguration*
• Externalization of labour and organizational restructuring. • A more highly qualified workforce but with different demands: 'commitment', 'putting the job first', presentism ('being there') and willingness to be mobile. • More non-linear careers – with periods of labour market disconnection/ investment in education capital through skills updating and skills enhancement; and standard employment contracts. • The centrality of Information Technology (IT) skills. • More remote information and communication technology (ICT)-based work: through use of laptops, fax machines, cellular telephones, etc. – 'enabling' individuals to work anytime, anywhere (including at home, while commuting, on business trips, etc.).	• An apparent expansion in the spatial horizons of managers and professionals for career development – often necessitating international mobility. • The blurring of business travel, short term business assignments and residential mobility. • The re-arrangement of the spatial and temporal linkages between home and work, with tasks for paid work undertaken in a variety of locations such as while travelling, whilst at home, and whilst socializing. • Residences increasingly chosen for their access to a number of labour markets (near to motorway hubs/airports); or willingness to be a dual location or commuter couple.

Source: Adapted from Stanworth (2000: 23) and Hardill (2002).

Work-related travel

Trends towards longer and more complex journeys

This section focuses on changes to commuting patterns, which for the highly skilled have become increasingly complex and difficult to 'manage'. A person's journey to work may cover a long distance and involve several modes of transport, such as a drive from home to the railway station, followed by a rail journey to the centre of a city, then a tram journey to the workplace. For some individuals journeys to work may not be undertaken on a daily basis, but rather may involve journeys to different locations at different times of day and of the week. However, it is worth remembering that only a minority (albeit an important minority) are undertaking such long-distance commutes; for most people, commutes still take place over short distances on a fairly regular basis (Green and Owen, 2006).

Over recent decades commutes have become longer, measured by distance covered, but less so by time taken. Journey-to-work distances have more than trebled since the end of the nineteenth century; over the period from the 1960s to the 1990s average distance to work increased from 10.2km to 14.6km.

However, the average commuting time has increased by just one minute per journey between the 1960s (33.5 minutes) and 1990s (34.5 minutes) (Pooley *et al.*, 2005: 114). There has also been a marked shift in mode of travel: 'The Twentieth Century has seen a major shift away from public transport and towards the private motor car' (Pooley and Turnbull, 1999: 287). For some individuals the 'choice' to use the car for the commute is a matter of preference, but for others it may not be a 'choice' at all because their work requires access to a car during the day, or because a car is needed to fulfil household responsibilities at either end of (or during) the working day, or because no suitable public transport is available from their residential location.

Lengthy commutes are particularly pronounced among managers and professionals who undertake approximately twice the amount of work-related travel as manual workers (Doyle and Nathan, 2001). Such lived experience has implications on individuals, particularly since travel can be stressful. Indeed, the results of a small-scale survey in Greater Nottingham (Wheatley, 2007) suggest that workers increasingly look upon their journey to work as a significant inconvenience, affecting the dynamic of the working day – as illustrated by the comment of a Nottingham commuter on the need to start early in order to acquire a car parking space: 'If you're not in by 8 am, forget it.'

Commuting as a substitute for migration

Employers' requirements for some staff to be spatially mobile have led some households with one or more managerial/professional workers to locate close to transport nodes, affording access to several labour markets, but resulting in more extensive and frequent work-related travel (Green, 1997). Such nodal locations may represent a compromise between partners' workplace locations (Doyle and Nathan, 2001: 9). This though may have significant impacts on the daily movements of household members, sometimes with both partners undertaking lengthy commutes – for example, one dual career couple from Wheatley's (2008) Nottingham survey recorded a 64 kilometres commute by one partner, and 96 kilometres by the other partner. Such 'nodal' locations and associated patterns of longer daily commutes, and sometimes weekly commutes and other types of non-permanent movement encapsulated in the term circulation (see Figure 16.1) mean that permanent residential migration to new workplace locations may be unnecessary (Green, 1997; Hardill and Green, 2003). In turn, this leads to a growth in mobile working.

Challenges for public policy

The location, commuting and migration decisions of households impact on public policy in a number of ways. Planning policy is increasingly required to acknowledge the importance of nodal living to many households, and some local authorities market themselves as nodal locations – for example, north Staffordshire has been promoted as a nodal location with a good transport infrastructure

<table>
<tr><td colspan="3" rowspan="2">Duration</td><td colspan="2">Reason for move</td></tr>
<tr><td>Production-related</td><td>Consumption-related</td></tr>
<tr><td rowspan="3">Time</td><td colspan="2">Permanent relocation</td><td>Labour migration (inter- and intra-organizational moves)</td><td>Housing adjustment
Amenity-led migration</td></tr>
<tr><td rowspan="2">Circulation</td><td>At least one (usually several) ovcrnight stay(s)</td><td>Short-term assignments

Long-distance (weekly) commuting</td><td>Holidays</td></tr>
<tr><td>Daily</td><td>Commuting</td><td>Shopping</td></tr>
</table>

Figure 16.1 Migration and circulation trade-offs (source: Hardill and Green 2003, 213).

affording access to labour markets in the metropolitan West Midlands and Greater Manchester. However, the preference of some households for semi-rural living and the location of some workplaces in out-of-town business parks, combined with 24/7 working practices, are not well served by public transport networks which are best suited to serving predictable medium to high volume flows on fixed routes. However, current transport policy in the UK places emphasis on reducing reliance on cars in order to ameliorate traffic congestion and improve the environment. Policy instruments currently include reducing the amount of car parking space, car park levies, congestion charging, park and rides, and improvements in public transport. But transport policy is not merely directed at changing commuting patterns; it also impacts upon the operation of businesses as business travel as part of the job is a growing feature of postmodern life. Policy must therefore reconsider current public transport networks and infrastructure, with new recognition for the changes in everyday mobility – in both 'work' and 'non-work' spheres. As noted above, a wide range of activities may be undertaken alongside the commute, including the 'school run' and shopping. Such 'trip chaining' causes complex issues of timetabling, rendering some people reliant on their cars during the working day. For example, an area manager employed by one Nottingham organization arrives at his workplace, but often leaves within an hour, subsequently visiting a number of other locations to attend meetings, and monitor the progress, before returning to his workplace an hour or so before the end of the working day to report back his findings (Wheatley, 2007).

Yet the information base on mobile work patterns is very poor. It is always the case that classifications used in information collection lag behind reality, but this is especially the case here. The UK Census of Population collects information on commuting patterns – assuming that individuals do the same commute every day to the same fixed location by the same mode of transport; the reality is complex spatial and temporal diaries in a 24/7 economy, which current statistics do not capture well. Hence, there is a limited evidence base for informing policy.

Work-space/place

ICT and the blurring of boundaries

Felstead *et al.* (2004) have argued that (paid) work today is an 'activity', not a 'place' (Felstead *et al.*, 2004), as work is 'leaking' out of organizations into more public domains such as cafes, hotel lobbies, etc. as well as into the home space. This section focuses on how mobile work practices are inextricably linked with the blurring of the home/work space.

ICT developments continue to impact on the nature, location and timing of paid work. Recent developments in ICT, especially in relation to mobile technologies, have included the increasing speed of broadband internet connections and the availability of wireless communications (mobile phones and wifi). ICT has had a revolutionary impact on work, providing virtual mobility, linking workplace locations and allowing 'direct on-line contact' (Castells and Aoyama, 1994: 26). However, impacts may be negative as well as positive. There is evidence that workloads and work intensity have increased (Perrons *et al.*, 2005) and that the working day has lengthened (Doyle and Reeves, 2001: 30). It has also been argued that the increased mobility offered by mobile technologies strengthens the hold economic activity has over individuals' lives (Jones, 1990: 254) – leading to a blurring of work and home lives.

Undertaking paid work in the home, or telework, whether on a regular or on an ad hoc basis, may provide greater flexibility for individuals in both the location and timing of work. It may offer the most appropriate solution for employees, particularly those in dual earner and dual career households, if they are to achieve a constructive balance between work and the rest of their lives (Hill *et al.*, 1996). However, the opportunity for such flexible working is not available to all employees, nor is it unproblematic. For instance, is there adequate physical space for working at home? Even with adequate space, further issues remain around the use of the home for work. Dividing not only space, but also time, for teleworking can be considerably challenging. There is perhaps a tendency for individuals to 'telework at all hours, interrupting meals, sleep, leisure and romance, in order to make time for their work' (José de Freitas Armstrong, 1999: 52). Research has indicated that, while some households may benefit from the greater flexibility offered by teleworking, others struggle with the blurring of the boundary between work and the home (Hardill *et al.*, 1997; Hill *et al.*, 1996).

Teleworking on a regular basis may also have significant career implications for workers, since in some instances careers are built on showing commitment in terms of giving time to work as described in 'presenteeism' (Erikson and Goldthorpe, 1992: 42). Workers whose office is in the home, may be disadvantaged by the 'invisibility' of work at home, and the social isolation of telework does not create opportunities for developing social networks in work (Sparrowe *et al.*, 2001).

Planning, workplaces and housing

Despite some of the changing work practices outlined above (and described in more detail in the next section) it remains the case that most paid work tasks are still executed in offices and workplaces within central urban areas or surrounding industrial parks (Breheny, 1999), while residences continue to be some distance from places of work.

On a large scale, the change in the use of residences, towards a location of paid work, may have significant impacts in relation to transport and planning. Current planning legislation actively discourages the use of the home as a location for paid work, yet current government advice on planning acknowledges 'mixed use' development, in which teleworking is encouraged as a means for reducing the requirement for regular work travel. However, this is currently found on an informal basis, with few examples of mixed use living and working space being planned (Dwelly, 2000).

Congestion may be removed from urban centres, but may increase elsewhere. Electronic communication does not substitute entirely for face-to-face communication (Britton *et al.*, 2004: 810–811). As noted above, the fact that individuals may be working at home on a relatively ad hoc basis for some days each week, attending their workplace during the remainder of the week, makes for complex patterns of movement, around which it is difficult to plan.

Policy implications for individuals, other household members and employers

The blurring of boundaries between 'work' and 'home' may have a number of significant impacts on individuals and other members of their household, as suggested above, and for employers.

For individuals there may be stress-related issues, including feeling as though they must always make themselves available, or that they are always on call (checking emails, telephone messages, etc.). There is also the potential isolation of working alone (Department for Trade and Industry, 2000), possibly resulting in the loss of identity and supportive social networks afforded by location at the workplace (Sparrowe *et al.*, 2001). A major cause of stress among teleworkers is likely to be a result of difficulties in dividing time and space between work and the home. The temptation for workers to check emails, or finish a report, may also impact on the family; the work of one household member invading the space and time of their partner and children.

Working from home raises health and safety issues for employers. They have a responsibility to ensure the health and safety of their workers under the Health and Safety at Work Act 1974. Employers must make a risk assessment of all work carried out by their employees whether at the workplace or at home, under the Management of Health and Safety at Work Regulations 1992 (Department for Trade and Industry, 2000: 26). Subsequently there may be issues of invasion of privacy for employees using their home as a location of work, as their employer must inspect the working environment and assess the risk (Fairweather, 1999). Evidence suggests that homeworkers do not receive the health and safety protection commensurate with their employment (Vassie, 2000: 541).

The factors described above can form significant barriers to the success of teleworking as an instrument for increasing flexibility, opportunity and the balance between work and other spheres of life. Employers' policy must acknowledge the need to be insured adequately to provide workers with cover for the potential adverse affects of working in the home. Beyond the health and social impacts of the blurring of home and work, further effects are felt in relation to the costs of implementing these schemes. While teleworking offers the opportunities for employers to perhaps reduce the size of their office space, set-up costs for teleworking can be considerable.

Mobile work practices

Developments in mobile technologies and the increasing prevalence of work

Developments in mobile technologies have broadened the potential location and timing of paid work. Employees are able to conduct business within the home via the internet and telephone networks, and on the move via mobile phones and PDAs, or combination devices, such as the BlackBerry. ICT is linking workplace locations (Castells and Aoyama, 1994: 26), and enabling work to be performed almost anywhere, with employees regularly engaging their laptops, PDAs and mobile phones to keep in touch with their workplace while visiting clients, or undertaking work travel, in airports, on trains or in the car. This may offer significant benefits in relation to multi-tasking and employees enhancing the productivity of time already allocated for work travel, where work would previously have been impossible. However, this time is unlikely to be recognized by employers as 'work time', and will therefore probably not be met with compensation in the form of pay, or time off in lieu.

Pooley *et al.* (2005: 2–3) suggest that the movements of individuals cannot simply be divided into different categories. Instead they propose that concepts of mobility are better understood as part of a 'mobility continuum'. The mobility continuum (Figure 16.2) details the range of household movements that comprise mobility, ranging from simple everyday movements, including the commute, work travel, relocation and virtual mobility.

Business travel includes irregular business trips, daily movements to a range of workplace locations, and short-term assignments to a particular location for a

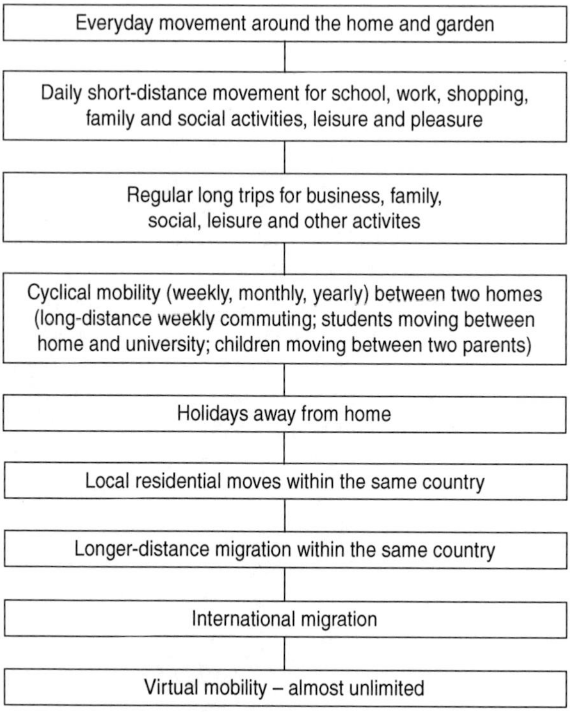

Figure 16.2 The mobility continuum.[1]

period of time; (here it is notable that employer policies for such assignments tend to be less generous than the terms of relocation packages (Green and Canny, 2004)). In addition the breadth of work travel is increasing, with employees likely to undertake travel not only to other locations within the city/region of their main workplace, but also inter-regional travel and overseas travel. Those working for multinational organizations may be asked to attend a meeting in Berlin on a Monday, and travel to New York to present a report on the Friday. While the costs of this travel will almost certainly be paid by the employer, the time taken for this type of frequent work travel is unlikely to be recovered through commensurate pay, and employers are also unlikely to appreciate the negative affects on the individual and their household, of a demanding travel schedule.

Mobile workers are consequently likely to be under greater time constraints, as they combine work time with work travel. Their workloads are likely to be increased through the improvements in mobile technologies (Perrons *et al.*, 2005), resulting in workers feeling that they are on call at all hours, similarly to those performing work in the home, postponing or interrupting other activities to complete tasks required by their paid employment (José de Freitas Armstrong,

1999: 52). Moreover, the increased mobility provided by mobile technologies may simply strengthen the hold economic activity has over individuals' lives, promoting further encroachment of work into locations and times previously free from work (Jones, 1990: 254).

Labour market restructuring, employment regulation and work–life balance

Changes in mobility are creating new opportunities for working. Consequently, measurement of when an individual is working has become increasingly difficult. Policies targeting patterns and conditions of work include monitoring and limiting the number of hours which can be worked, and promoting balance between work and other spheres of life. Many of these policies are designed and implemented nationally, and in the case of EU members, including the UK, there are also EU directives, such as the European Working Time Directive (EWTD), implemented in 1998 in the UK as the Working Time Regulations (WTR). These legislative acts are designed to protect against the possible negative effects which long working hours have on individuals' health (European Commission, 2005). However, these policies were not designed with workers who undertake working activity, during travel, in mind.

Policy on regulation of working time may be developed transnationally, but is implemented nationally. Hence, multinational companies face new policy challenges associated with changing working practices, such as offshoring of workplaces (e.g. banking and finance call centres). Perhaps the greatest difficulties are experienced by the increasing number of global employers, who must tailor their workplace policy for their workers in a range of different policy environments (Adler, 1994, 1997).

Mention has already been made of some of the health and safety challenges posed by mobile working and more general spatially diffused working practices. Health and safety policy is implemented nationally, with the responsibility of enforcement left to employers. However, questions remain over the level of health and safety monitoring by employers outside conventional workplaces. Moreover, working time regulation and work–life balance policy needs to pay greater cognizance to the increasing fluidity of work, taking place not only in the workplace, but in the home, and on the move.

Home–work dynamics

Complex choreographies

Reference has been made in previous sections to the way in which changing work practices, including mobile working, blur the boundaries between 'home' and 'work'. Drawing on the evidence presented above, this section concentrates on discussion of the complexities of the lived reality of increasing mobility on households. For those couple households where both partners have a deep commitment to the labour market, combining successful dual careers, and achieving balance between work and life, may be particularly challenging. While some households may experience significant benefits from the greater flexibility offered by teleworking, others struggle with the blurring of the boundary between work and the home (Hill *et al.*, 1996).

Managing two careers and juggling work schedules can make life complicated (Carnoy, 2000: 116). In households where children are present, couples may split the responsibilities of taking their children to school and picking them up at the end of the school day, but they may not (Hardill, 2002). Flexible working arrangements may provide flexibility to juggle domestic responsibilities, such as the school run, but this may be at the expense of being a two-car household against a policy background of trying to reduce car dependency.

Achieving balance?

The work–life balance campaign aims to make employers more aware of the need for employees to obtain a balance between work and life. To help employees achieve a successful balance, Nottingham organizations surveyed by Wheatley (2008) offered a number of forms of flexible working arrangements including part-time working, compressed hours, flexi-time, and teleworking or home–working, but sometimes these were in conflict with other work commitments or with preferences in non-work spheres. For example, some respondents indicated a conflict between accessing flexi-time, and accessing limited car parking at work. A number of employees expressed that they felt constrained to arrive at their workplace early in order to obtain car parking for the day, thus reducing the flexibility in their hours of work. Standard working activities, such as attendance at meetings, and meeting fixed deadlines, also form a barrier to working non-standard hours, which some employees preferred in order to fit in non-work aspects of their daily routines.

The success of flexible working schemes is very much dependent on the employer embracing the need to improve employees' work–life balance, acknowledging the growing complexity of household routines. Managers and professionals are likely to be 'work-rich', 'time-poor'. This is especially likely to be the case for highly skilled female workers, who tend to undertake significantly greater responsibilities within the household than their male counterparts (McDowell *et al.*, 2005). However, managers and professionals may benefit more from greater latitude for flexible working than, say, those in lower-skilled occupations, and this greater autonomy may in many ways be advantageous to them.

Conclusion: implications for public policy

This chapter has explored the impacts and policy implications of changes in mobility in the context of the workplace and the household. It has shown how mobility is impacting on work-related travel, the blurring of the home and work, mobile working practices, and on home–work dynamics.

Work travel is a significant part of many workers' routines. This time is used to travel to work, but also for work, engaging mobile technologies – e.g. wifi. Working now takes place not only in the workplace, but increasingly on the move due to the development of mobile technologies. The 24/7 economy is having significant impacts on the household, with partners working weekends, or embarking on work travel requiring week-long visits abroad. This has particular policy implications in relation to the Working Time Regulations, as new questions arise of 'When are workers working?' 'What business travel is and isn't included?' And whether the daily commute is now a 'place of work' – at least for some people? Challenges are also faced by global employers, who are located in a number of national regulatory environments, and who ask their employees to undertake regular large-scale work travel. These employers must design and implement workplace policy while obeying the legislation and regulation of a number of different countries.

The home is now a place of paid work as well as unpaid work. Teleworking may yield significant benefits to the employee in managing their household routines due to increased flexibility. However, those working in the home face a number of difficulties, including divisions of space and time for work, stress caused by isolation, and the possible encroachment of work into other household activities, resulting in workers feeling that they are always 'on call' and/or 'at work'. There are important issues here about what exactly constitutes the 'non-work' space of the employee and whether, and how far, employers should intrude into it. Moreover, the working routines of one household member may therefore impact greatly on the personal lives of others. Flexible working arrangements can and do help employees to juggle the increasingly blurred boundary between home and work. A number of significant benefits may be felt by the household when these arrangements are employed, however policy must acknowledge the impacts on all household members.

The evidence and discussion presented in this chapter have highlighted the complexity of new dynamics in the practices and location of work. There is a need to improve the evidence base to help inform policy: for economic regeneration and development; for planning; for housing and; for health and safety, for work–life balance, etc. Currently, the paucity of the information base means that policy makers only have a relatively poor grasp of the scale and nature of mobile working. It is not well captured by conventional analyses of commuting and working patterns. Moreover, by its very nature it is a dynamic phenomenon and one that is subject to change over time. Hence, there is a pressing need for improved data collection instruments and revisions to surveys to collect information that will help monitor developments and establish a more robust evidence base for planning.

Note

1 The geographical scale of mobility increases successively from the top to the bottom of the diagram.

References

Adler, N. (1994) 'Competitive Frontiers: Women Managing across borders', *Journal of Management Development*, 13: 24–41.

Breheny, M. (1999) (ed.) *The People: Where Will They Work?* London: Town and Country Planning Association.

Britton, N. J., Halfpenny, P., Devine, F. and Mellor, R. (2004) 'The Future of Regional Cities in the Information Age: The Impact of Information Technology on Manchester's Financial and Business Services Sector', *Sociology*, 38, 4: 795–814.

Carnoy, M. (2000) *Sustaining the New Economy*. Cambridge, MA: Harvard University Press.

Castells, M. and Aoyama, Y. (1994) 'Paths towards the Informational Society: Employment Structure in G-7 countries 1920–90', *International Labor Review*, 1: 133.

Department of Trade and Industry (2000) *Working Anywhere: Exploring Telework for Individuals and Organisations*, 2nd edition. London: UK Online for Business.

Doyle, J. and Nathan, M. (2001) *Wherever Next: Work in a Mobile World*, the Industrial Society. Loughton, Essex: CW Press.

Doyle, J. and Reeves, R. (2001) *Time Out: The Case for Time Sovereignty*, the Industrial Society. Loughton, Essex: CW Print.

Dwelly, T. (2000) *Living at Work: A New Policy Framework for Modern Home Workers*. York: Joseph Rowntree Foundation.

Erikson, R. and Goldthorpe, J. H. (1992) *The Constant Flux: A Study of Class Mobility in Industrial Societies*. Oxford: Clarendon Press.

European Commission (2005) Working Time Directive. Available at: www.cec.org.uk/info/pubs/bbriefs/bb28.htm.

Fairweather, N. B. (1999) 'Surveillance in Employment: The Case of Teleworking', *Journal of Business Ethics*, 22, 1: 39–49.

Felstead, A, N., Jewson, N. and Walters, S. (2004) *Changing Places of Work*. London: Palgrave.

Green, A. E. (1997) 'A Question of Compromise? Case Study Evidence on the Location and Mobility Strategies of Dual Career Households', *Regional Studies*, 31, 7: 643–659.

Green, A E. and Canny, A. (2003) *Geographical Mobility: Family Impacts*. Bristol: Policy Press.

Green, A. E. and Owen, D. W. (2006) *The Geography of Poor Skills and Access to Work*. York: Joseph Rowntree Foundation.

Hardill, I. (2002) *Gender, Migration and the Dual Career Household*, International Studies of Women and Place series. London: Routledge.

Hardill, I. and Green, A. (2003) 'Remote Working: Altering the Spatial Contours of Work and Home in the New Economy', *New Technology, Work and Employment*, 18: 3.

Hardill, I., Green, A. E. and Dudleston, A. C. (1997) 'The "Blurring of Boundaries" between "Work" and "Home": Perspectives from Case Studies in the East Midlands', *Area*, 29, 4: 335–343.

Hill, E. J., Hawkins, A. and Miller, B. (1996) 'Work and Family in the Virtual Office: Perceived Influences of Mobile Telework', *Family Relations*, 45, 3: 293–301.

Jones, B. (1990) *Sleepers, Wake!* Melbourne: Oxford University Press.

José de Freitas Armstrong, N. (1999) 'Flexible Work in the Virtual Workplace: Discourses and Implications of Teleworking', in A. N. Felstead and N. Jewson (eds) *Global Trends in Flexible Labour*. London: Macmillan Press.

McDowell, L., Perrons, D., Fagan, C., Ray, K. and Ward, K. (2005) 'The Contradictions and Intersections of Class and Gender in a Global City: Placing Working Women's Lives on the Research Agenda', *Environment & Planning A*, March, 37, 3: 441–461.

Perrons, D., Fagan, C., McDowell, L., Ray, K. and Ward, K. (2005) 'Work, Life and Time in the New Economy', *Time and Society*, 14, 1: 51–64.

Pooley, C. G. and Turnbull, J. (1999) 'The Journey to Work: A Century of Change', *Area*, 31, 3: 281–292.

Pooley, C. G., Turnbull, J. and Adams, M. (2005) *A Mobile Century? Changes in Everyday Mobility in Britain in the Twentieth Century*. Ashgate Publishing.

Rapoport, R. and Rapoport, R. (1976) *Dual Career Families Re-examined*. London: Martin Robertson.

Sparrowe, R. T., Liden, R. C., Wayne, S. J. and Kraimer, M. L. (2001) 'Social Networks and the Performance of Individuals and Groups', *Academy of Management Journal*, 44, 2: 316–325.

Stanworth, C. (2000) 'Women and Work in the Information Age', *Gender, Work and Organization*, 7, 1: 20–32.

Vassie, L. (2000) 'Managing Homeworking: Heath and Safety Responsibilities', *Employee Relations*, 22, 6: 540–554.

Wheatley, D. (2008) 'Managing' Complex Commuting and Working Patterns in the 21st Century: The Case of Nottingham', *Regions*, 269 (Spring): 6–9.

Index

Page numbers in *italics* indicate tables; pages numbers in **bold** indicate figures.